Impact of Climate Change on Livestock Health and Production

About the Editors

Gangadhar Nayak is presently working as Professor and Head, Department of Animal Breeding and Genetics in College of Veterinary Science and Animal Husbandry, Odisha University of Agriculture and Technology (OUAT), Bhubaneswar, Odisha. He has published 02 books and more than 95 research papers in national and international journals.

Kautuk Kumar Sardar is presently working as Professor of Pharmacology & Toxicology at the Odisha University of Agriculture & Technology, Bhubaneswar. He is MVSc Gold Medalist from OUAT and Ph.D. from ICAR-Indian Veterinary Research Institute, Izatnagar, Bareilly, UP with brilliant academic record. He is an eminent pharmacologist with over 28 years of vast experience in teaching, research and extension.

Dr. Sardar has published more than 100 research papers in journals of National and International repute with publication of four manuals; five compendia; over 100 abstracts. He has attended more than 34 National Conferences in India; delivered several lecture-cum-practical demonstrations including 12 invited lectures at different trainings, workshops, seminars and conferences; adjudged as Member, editorial board and referee to various International and National Journals of repute; acted as Course Co-ordinator of ICAR-Winter School; delivered many TV and radio talks on various issues related with livestock health and production. He has been conferred with "National Alarsin Award (Indian Veterinary Association - 2000); Member, National Academy of Veterinary Sciences, India (2005); Member, National Academy of Sciences, India (2009); Member, National Academy of Biological Sciences, India (2012); Indian Society of Veterinary Medicine Appreciation Award (2012) as well as IAAVR - Dr. R. M. Sojatia Memorial Award (2016). Under his able guidance, more than 20 scholars have successfully completed their Masters' and PhD Degree in Veterinary Pharmacology and Toxicology as well as in other disciplines of Veterinary and Animal Sciences.

Bhabesh Chandra Das is presently working as Assistant Professor in College of Veterinary Science and Animal Husbandry, Odisha University of Agriculture and Technology (OUAT), Bhubaneswar, Odisha. He has published 09 books and more than 108 articles in the national and international journals.

Debiprasanna Das is presently working as Assistant Professor in College of Veterinary Science and Animal Husbandry, Odisha University of Agriculture and Technology (OUAT), Bhubaneswar, Odisha. He was University topper at Master level and awarded with 06 Gold medals from TANUVAS. He was also recipient of INSPIRE fellowship of Department of Science and Technology, Government of India for his Ph.D degree. He is presently an Executive Member of Indian Association of Veterinary Pathology(IAVP). He has published more than 47 articles in the national and international journals.

Impact of Climate Change on Livestock Health and Production

G D Nayak
Professor and Head
Department of Animal Breeding and Genetics
College of Veterinary Science and Animal Husbandry Odisha
University of Agriculture and Technology Bhubaneswar – 751 003,
Odisha, India

K K Sardar
Professor
Department of Pharmacology and Toxicology
College of Veterinary Science and Animal Husbandry Odisha
University of Agriculture and Technology Bhubaneswar – 751 003,
Odisha, India

B C Das
Assistant Professor
Department of Veterinary and Animal Husbandry Extension College
of Veterinary Science and Animal Husbandry Odisha University of
Agriculture and Technology Bhubaneswar – 751 003, Odisha, India

D P Das
Assistant Professor
Department of Veterinary Pathology
College of Veterinary Science and Animal Husbandry Odisha
University of Agriculture and Technology Bhubaneswar – 751 003,
Odisha, India

CRC Press is an imprint of the
Taylor & Francis Group, an **informa** business

NEW INDIA PUBLISHING AGENCY
New Delhi-110 034

First edition published 2023
by CRC Press
4 Park Square, Milton Park, Abingdon, Oxon, OX14 4RN

and by CRC Press
6000 Broken Sound Parkway NW, Suite 300, Boca Raton, FL 33487-2742

CRC Press is an imprint of Informa UK Limited

British Library Cataloguing-in-Publication Data
A catalogue record for this book is available from the British Library

ISBN: 9781032428710 (hbk)
ISBN: 9781032428727 (pbk)
ISBN: 9781003364689 (ebk)

DOI: 10.4324/9781003364689

Typeset in Times New Roman
by NIPA, Delhi

भारतीय कृषि अनुसंधान संस्थान
कृषि भवन, डॉ राजेंद्र प्रसाद रोड
नई दिल्ली – 110001
INDIAN COUNCIL OF AGRICULTURAL RESEARCH
Krishi Bhawan, Dr. Rajendra Prasad Road
New Delhi-110001

Dr. Bhupendra Nath Tripathi, PhD
DICVP, FNAAS, FNAVS, FIAVP
Deputy Director General (Animal Science)

Foreword

The livestock sector significantly contributes to the livelihoods of one billion poor population in the world and provides employment to about 1.1 billion people. There is a growing demand for livestock products, and its rapid growth in developing countries has led to "livestock revolution". Worldwide milk production is expected to increase from 664 million tons (in 2006) to 1077 million tons (by 2050), and meat production will double from 258 to 455 million tons during the same period. As such, livestock products are important agricultural commodities for global food security because they provide 17% of global energy consumption and 33% of global protein consumption.

Due to increasing global temperature, climate shift and shrinkage of basic resources like land, water and feed, the livestock and poultry production is facing stiff challenges particularly when demand for livestock products is expected to increase by 100% by mid of the 21st century. Climate change affects livestock production directly through biotic and abiotic stresses, which manifest in the form of disease outbreaks, introduction of new diseases, reduction or loss of production, reproductive disorders, physiological disturbances, etc. Indirectly, climate change affects through impeding the maximization of agricultural and fodder production, loss of biodiversity, spread of vector-borne diseases, food-borne diseases, etc. Further, the increasing ambient temperature also changes pathogens to become thermo-tolerant and retain or increase their pathogenicity even at higher temperatures, and reduces the genetic resistance of livestock, thus making them more susceptible to diseases.

The livestock sector contributes 14.5% of global greenhouse gas (GHG) emissions, and thus may increase land degradation, air and water pollution and may eliminate 15% - 37% of all species in the world resulting in loss of biodiversity. Therefore, if livestock numbers continue to increase and feeding practices are not changed, GHG emissions due to livestock production will also increase. There are several climate change adaptation and mitigation recommendations, where adaptation strategies can improve the resilience of crop and livestock productivity to climate change. Mitigation measures could significantly reduce the impact of climate change on livestock. Adaptation measures involve modifications in production and management system, breeding strategies, institutional and policy changes, science and technological advances, and changing the farmers' perception and adaptive capacity.

In this compilation, authors have meticulously brought forward how both climate change and livestock sector can reciprocally impact each other and mitigation strategies to significantly reduce the adverse impact of one on another with aim of developing a climate resilient livestock and poultry production system in the country. I am confident that this book will act as a reference in the field of climate change and livestock production.

-sd/-
Bhupendra Nath Tripathi

Preface

Climate change is one of the important challenges facing the world today. It is no more an environmental concern but emerged as the biggest developmental challenge for animal-agriculture system. The consequences of climate change manifest themselves in multiple ways, including increased variability and intensity of extreme weather events leading to unpredicted production. The search for solutions to mitigate climate change and to adapt to its consequences is urgent need of the hour. Periodic evaluation of different approaches adopted to mitigate the adverse impact of climate changes on agriculture and allied sectors needs scientific validations to provide inputs to policy makers to design and develop programmes suitable for livestock production system. Climate change affects livestock production system and thereby affecting food security of the country. Present climate variability is negatively impacting livestock production due to untimely unset of diseases, lack of availability of water, deforestation, shrinkage of pasture land, etc.

Climate change adaptation and mitigation strategies are critical to protect livestock production and sustainable husbandry practices. Diversification of livestock-agriculture practices using different crop varieties and integrating mixed crop livestock system may be one of the promising adaptation measures. There is growing interest among the researchers, policy makers and other stake holders in understanding the interactions of climate change and livestock production. Very limited researches regarding the impacts of climate change on livestock production is available.

The present book on "Impact of Climate Change on Livestock Health and Production" contains 30 chapters contributed by the learned authors of national and international repute covering on various latest aspects involving diversification of livestock and crops, integration of livestock systems with forestry and crop production, drought and heat wave tolerant varieties, strategies for reduction of Green House Gases emission from ruminants, application of GIS and remote sensing technologies, breeds with inherent genetic capabilities to adapt to climate change. etc. This book also emphasises the climate change adaptation, mitigation practices, and policy frameworks for promotion of sustainable livestock and poultry production.

Information given in the book is based on knowledge and understanding of the subjects of various experts. However, the editors are not responsible for any typographical errors or other mistakes in the book. The editors greatly acknowledge the contributions made by the authors of the relevant chapters. The contributors of the concerned chapters are solely responsible for the originality of their contents.

The editors are highly thankful to Dr. R.K Mahapatra, Former Chief Librarian, Central Library, Odisha University of Agriculture and Technology Bhubaneswar for his constant support and guidance for publication of the book.

We are immensely grateful and indebted to Prof. Pawan Kumar Agrawal Hon'ble Vice-Chancellor, Odisha University of Agriculture and Technology Bhubaneswar for his constant inspiration, unflinching encouragement precious guidance and moral support for preparation of this book.

G D Nayak
K K Sardar
B C Das
D P Das

Contents

List of Contributors

A. K. Kundu, Professor & Head, Department of Animal Physiology, College of Veterinary Science and Animal Husbandry, Odisha University of Agriculture and Technology, Bhubaneswar – 751 003, Odisha, India

A.K. Panda, Principal Scientist, ICAR-Central Institute for Woman in Agriculture, Bhubaneswar-751 003, Odisha, India

A. Sahoo, Director, ICAR-Central Sheep and Wool Research Institute, Avikanagar – 304 501 Rajasthan, India

Amit Kumar, ICAR-Indian Veterinary Research Institute, Izatnagar-243 122, Bareilly, Uttar Pradesh, India

B.C. Das, Assistant Professor, Department of Veterinary and Animal Husbandry Extension, College of Veterinary Science and Animal Husbandry, Odisha University of Agriculture and Technology, Bhubaneswar – 751 003, Odisha, India

B.P. Mishra, Joint Director (Research), ICAR-Indian Veterinary Research Institute, Izatnagar-243122, Bareilly, Uttar Pradesh, India

B.S. Rath, Professor & Head, Department of Agricultural Meteorology, College of Agriculture, Odisha University of Agriculture and Technology, Bhubaneswar – 751 003 Odisha, India

B. Sajjanar, ICAR-Indian Veterinary Research Institute, Izatnagar-243 122, Bareilly, Uttar Pradesh, India

Basudev Behera, Professor & Head, Department of Agronomy, College of Agriculture Odisha University of Agriculture & Technology, Bhubaneswar – 751 003, Odisha, India

C. Balachandran, Vice-Chancellor, Tamil Nadu Veterinary and Animal Sciences University (TANUVAS), Madhavaram Milk Colony, Chennai - 600 051, Tamil Nadu, India

Chinmoy Mishra, Assistant Professor, Department of Animal Breeding and Genetics, College of Veterinary Science and Animal Husbandry, Odisha University of Agriculture and Technology, Bhubaneswar – 751 003, Odisha, India

D.B.V. Ramana, Principal Scientist (LPM), ICAR-Central Research Institute for Dryland Agriculture (CRIDA), Santoshnagar, Hyderabad – 500 0 59, Telangana, India

D.K. Karna, Associate Professor, Department of Animal Breeding and Genetics, College of Veterinary Science and Animal Husbandry, Odisha University of Agriculture and Technology, Bhubaneswar – 751 003, Odisha, India

D.P. Das, Assistant Professor, Department of Veterinary Pathology, College of Veterinary Science and Animal Husbandry, Odisha University of Agriculture and Technology Bhubaneswar – 751 003, Odisha, India

G.D. Nayak, Professor and Head, Department of Animal Breeding and Genetics, College of Veterinary Science and Animal Husbandry, Odisha University of Agriculture and Technology, Bhubaneswar – 751 003, Odisha, India

Gopal Krushna Tripathy, Deputy Director, Veterinary Officers' Training Institute Directorate of Animal Husbandry & Veterinary Services, Government of Odisha Bhubaneswar 751 006, Odisha, India

H.N. Malik, Scientist (Animal Science), Krishi Vigyan Kendra, Bhawanipatna, Odisha University of Agriculture and Technology, Bhubaneswar – 751 003, Odisha, India

J. Mohanty, Principal Scientist, ICAR-Central Institute of Freshwater Aquaculture Kausalyaganga, Bhubaneswar – 751 002, Odisha, India

K.K. Sardar, Professor, Department of Pharmacology and Toxicology, College of Veterinary Science and Animal Husbandry, Odisha University of Agriculture and Technology Bhubaneswar – 751 003, Odisha, India

Kamdev Sethy, Assistant Professor, Department of Animal Nutrition, Odisha University of Agriculture and Technology, Bhubaneshwar – 751 003, Odisha, India

L. Samal, Assistant Professor, PG Department of Poultry Science, College of Veterinary Science and Animal Husbandry, Odisha University of Agriculture and Technology Bhubaneswar -751003, Odisha, India

M.K. Padhi, Principal Scientist, Regional centre, ICAR-Central Avian Research Institute Bhubaneswar-751 003, Odisha, India

Mohan R. Badhe, ICAR-Central Institute of Freshwater Aquaculture, Kausalyaganga Bhubaneswar – 751 002, Odisha, India

Mukesh Bhakat, Artificial Breeding Research Center, ICAR-National Dairy Research Institute, Karnal – 132 001, Haryana, India

N. C. Behura, Professor, PG Department of Poultry Science, College of Veterinary Science and Animal Husbandry, Odisha University of Agriculture and Technology Bhubaneswar - 751 003, Odisha, India

P.C. Mishra, Associate Professor and Head, Department. of Animal Reproduction Gynaecology & Obstetrics, College of Veterinary Science and Animal Husbandry Odisha University of Agriculture and Technology, Bhubaneswar– 751003, Odisha, India

P. K. Roul, Dean, Extension Education, Odisha University of Agriculture and Technology Bhubaneswar -751003, Odisha, India

Pravas Ranjan Sahoo, Assistant Professor, Department of Veterinary Biochemistry, College of Veterinary Science and Animal Husbandry, Odisha University of Agriculture and Technology, Bhubaneswar – 751 003, Odisha, India

Priyabrat Swain, Principal Scientist, ICAR-Central Institute of Freshwater Aquaculture, Kausalyaganga, Bhubaneswar - 751 002, Odisha, India

R. K. Swain, Professor, Department of Animal Nutrition, Odisha University of Agriculture and Technology, Bhubaneshwar – 751 003, Odisha, India

S. C. Sahu, Director, Centre for Environment & Climate, Siksha 'O' Anusandhan (Deemed to be University), Khandagiri, Bhubaneswar - 751 030, Odisha, India

S. K. Dash, Professor, Department of Animal Breeding and Genetics, College of Veterinary Science and Animal Husbandry, Odisha University of Agriculture and Technology, Bhubaneswar -751003, Odisha, India

S. K. Mishra, Professor, Department of Animal Nutrition, Odisha University of Agriculture and Technology, Bhubaneshwar – 751 003, Odisha, India

S. R. Mishra, Assistant Professor, Department of Animal Physiology, College of Veterinary Science and Animal Husbandry, Odisha University of Agriculture and Technology Bhubaneswar – 751 003, Odisha, India

S. Vasantha Kumar, Associate Professor, Department of Livestock Production Management, Veterinary College and Research Institute, Tamil Nadu Veterinary and Animal Sciences University (TANUVAS), Tirunelveli - 627 358, Tamil Nadu, India

Sanat Mishra, Senior Scientist, Directorate of Planning, Monitoring & Evaluation, Odisha University of Agriculture & Technology, Bhubaneswar – 751 003, Odisha, India

Sandeep Pattnaik, Assistant Professor, School of Earth Ocean and Climate Sciences Indian Institute of Technology, Bhubaneswar, Argul, Jatni – 752 050, Odisha, India

Swagat Mohapatra, Assistant Professor, Department of Animal Physiology, College of Veterinary Science and Animal Husbandry, Odisha University of Agriculture and Technology, Bhubaneswar – 751 003, Odisha, India

T. K. Mohanty, Principal Scientist & In-Charge, Artificial Breeding Research Center ICAR-National Dairy Research Institute, Karnal – 132 001, Haryana, India

1

Impact of Climate Change on Livestock Health and Production

C Balachandran

Introduction

According to the Intergovernmental Panel on Climate Change, climate change refers to the long-term changes (typically decades or longer) in the average state of the climate with statistically significant variations.

The climate change, defined as the long-term imbalance of customary weather conditions such as temperature, radiation, wind and rainfall characteristics of a particular region, is likely to be one of the main challenges for mankind during the present century.

A general definition of climate change is a change in the statistical properties (principally the man and spread) of the climate system when considered over long periods.

Climate resilience

Climate resilience can be generally defined as the capacity for a socio-ecological system to: Absorb stresses and maintain function in the face of external stresses imposed upon it by climate change. Adapt, reorganize, and evolve into more desirable configurations that improve the sustainability of the system, leaving it better prepared for future climate change impacts.

Preamble

The earth's climate has warmed in the last century (0.74 ± 0.18°C) with the 1990s and 2000s being the warmest on instrumental record (Intergovernmental Panel on Climate Change-IPCC, 2007). Furthermore, the earth's climate has been predicted to change continuously at rates unprecedented in recent human history. The IPCC is an UN panel which deals with the science related to climate change. Current climate models indicated an increase in temperature by 0.2°C per decade and predicted that the increase in global average surface temperature would be between 1.8°C to 4.0°C by 2100 (IPCC, 2007). The temperature rises are much greater than those seen during the last century, when average temperatures rose only 0.06°C (0.12°F) per decade (National Oceanic and Atmospheric Administration, 2007). National Aeronautics and Space Administration (NASA) announced the five warmest years since the 1890s were 1998, 2002, 2003, 2004, and 2005. IPCC's latest report warns that climate change could lead to some impacts that are abrupt or irreversible.

Global climate change is primarily caused by greenhouse gas-GHG (carbon dioxide-CO_2, methane-CH_4 and nitrous oxide-N_2O) emissions that result in warming of the atmosphere (IPCC, 2013). Global warming and ozone layer depletion due to increased emission of GHG in the atmosphere have drawn world-wide attention with an alarming stage of iceberg melting, increased ocean level, local and global eco-system upsets, changes in the rainfall patterns, changes in pathogenesis of plants, animals and human beings and alteration in life of the people (Kumar *et al.*, 2008).

Human population is expected to increase from 7.2 to 9.6 billion by 2050 (UN, 2013). This represents a population increase of 33 per cent, but as the global standard of living increases, demand for agricultural products will increase by about 70 per cent in the same period (FAO, 2009a). Meanwhile, total global cultivated land area has not changed since 1991 (O'Mara, 2012), reflecting increased productivity and intensification efforts.

Estimated demand for animal-derived food in 2050 could be 70% higher than 2005 levels, i.e. beef and pork increase by 66 and 43%, respectively (Alexandratos and Bruinsma, 2012), poultry meat at 121% growth, eggs potentially 65% (Mottet and Tempio, 2017). This will lead to increase in livestock and poultry population. The capture fisheries is likely to remain same as fully harvested. Ocean acidification from cumulative GHG emissions is leading to biodiversity loss and threatening single-cell phytoplankton, the base of marine food chains accounting for >50%. Eutrophication and algal blooms from the runoff of nitrogen and phosphorous-based fertilizers into running water bodies are further depleting aquatic plant and animal species, food source for humans (Carpenter *et al.*, 1998; Canfield *et al.*, 2010). Aquaculture will dominate growth in the fish sector. Fishery production is expected to expand by >30 metric tons by 2030 and 95% will come from developing countries.

Human-created processes like deforestation, shifting cultivation, wasting and polluting water, decreasing agricultural land, increasing mining and industrial activities and spreading of urban areas with increasing population and indiscriminate use of land and water worsened the situation (Shepali, 2019).

Livestock products are an important agricultural commodity for global food security because they provide 17% of global kilocalorie consumption and 33% of global protein consumption (Rosegrant *et al.,* 2009). The livestock sector contributes to the livelihoods of one billion of the poorest population in the world and employs close to 1.1 billion people (Hurst *et al.,* 2005). Worldwide milk and meat productions are expected to increase 1077 and 455 million tons respectively by 2050 (Alexandratos and Bruinsma, 2012). At the same time, climate change will affect livestock production through competition for natural resources, quantity and quality of feeds, livestock diseases, heat stress and biodiversity loss.

Malnutrition due to lack of micronutrients, which are bioavailable predominantly in animal-source foods, lead to chronic diseases and hunger and hidden hunger remain problematic in the developing world. Overconsumption of meat is detrimental to health in the developed world, whereas on the reverse poorest as the under consumption of animal source food is detrimental to health (James and Palmer, 2015; Beal *et al*., 2017). Currently 86% of global livestock feed is made of materials that are not consumed by humans, and ruminants play an important role in that they are uniquely able to convert human-inedible forages (e.g., leaves and grass) into high-quality protein and a variety of micronutrients (Animal Agriculture, 2019).

Climate Change and India

Being homeothermic organisms, livestock must regulate their body temperature within a relatively narrow range to remain healthy and productive. However, high relative humidity reduces the effectiveness of evaporative cooling and during hot, humid summer weather the animal cannot eliminate sufficient body heat and body temperature rises (Smita and Michaelowa, 2007).

"Climate change is now affecting every country disrupting their economies affecting lives, costing people, communities and countries clearly today and even more tomorrow. Significant impacts of climate change include changing weather patterns, rising sea level and more extreme weather events. The poorest and most vulnerable people are being affected the most."

– Goal 13 UN Sustainable Development Goals.

Indian Livestock and Poultry Vs GHG Emissions

In India, although the emission rate per animal is much lower than the developed countries, due to huge livestock population, the total annual CH_4 emission is about 9-10 Tg from enteric fermentation and animal wastes. India possesses the largest livestock population in the world and accounts for the largest number of cattle (world share 16.1%), buffaloes (57.9%), second largest number of goats (16.7%) and third highest number of sheep (5.7%) in the world (FAOSTAT). Of the various livestock enterprises, dairying is most popular in the country and account for nearly 60% of these enteric emissions. Poultry production generates hatchery waste, litter (bedding material, saw dust, wood shavings and peanut hulls), offal, processing water and bio-solids. The major environmental pollutants are micro environment gases / pollutants like CO_2, CH_4, NH_4, and nitrous oxide.

Impact of Climate Change on Biodiversity

Climate change may eliminate 15% to 37% of all species in the world (Thomas *et al*., 2004). The IPCC (2014) Fifth Assessment Report states that an increase of 2 to 3°C above pre-industrial levels may result in 20 to 30% of biodiversity loss of plants and animals. By 2000, 16% of livestock breeds (ass, water buffalo, cattle, goat, pig, sheep, and horse) were lost (Thornton *et al.,* 2009). In addition, the FAO (2007) has stated that from 7,616 livestock breeds reported, 20% were at risk, and almost one breed per month was being extinguished. Cattle had the highest number of extinct breeds (N = 209) of all species evaluated. The livestock species that had the highest percentages of risk of breed elimination were chicken (33% of breeds), pigs (18% of breeds), and cattle (16% of breeds). However, the breeds at risk depends on the region. These breeds and species cannot be replaced naturally; therefore, future work that studies the inherent genetic capabilities of different breeds and identifies those that can better adapt to climate conditions is vital.

There is growing interest in understanding the interaction of climate change and agricultural production (Aydinalp and Cresser, 2008). There is still limited research regarding the impacts of climate change on livestock production (IPCC, 2014).

Impact of Climate Change on Livestock Production

Warm and humid conditions cause heat stress, which affects behavior and metabolic variations on livestock or even mortality. Heat stress impacts can be categorized into feed nutrient intake/utilization, animal production, reproduction, health and mortality (Table 1).

Table 1: Impact of climate change on livestock production

Impact of climate change on livestock production				
Growth	Milk production	Reproduction	Adaptation	Disease occurrence
Decrease in body weight	Decrease in milk production	Decrease in estrus expression	Increase in respiration rate	Increase in vector borne diseases
Decrease in body condition scoring	Decrease in milk quality	Decrease in fertilization	Increase in rectal temperature	Increase in trypanosomiasis
Decrease in average daily gain	Decrease in lactose	Decrease in conception rate	Increase in cortisol	Increase in bovine viral diarrhea
Decrease in feed intake	Decrease in milk fat	Decrease in folliculogenesis	Increase in heat shock proteins (HSP)	Increase in Rift valley fever
Decrease in feed conversion ratio	Decrease in SNF	Decrease in spermatogenesis	Increase in thyroid hormones	Increase in theileriosis
Decrease in allometric measurements	Decrease in milk protein	Decrease in testosterone	Increase in water intake	Increase in parasitic diseases

Reduced production and severe economic losses (Sejian *et al.*, 2016)

Impact of Climate Change on Livestock

About 17.5% of the dairy cattle in the country are now crossbred. Average annual temperature of our country is 25°C or higher. Upper temperature limit of comfort zone for maximum milk production is 27°C, about two degrees higher than the same reported in temperate countries. However, the average annual temperature is higher than this upper critical limit in several parts of the country, particularly southeastern region of Andhra Pradesh and Tamil Nadu. A likely consequence would be reduction of the total area where high yielding dairy cattle can be economically reared. Due to GHG emissions the likely range of increase in global average surface temperature by 2100, is predicted to be between 0.3 °C and 4.8 °C (IPCC, 2013). The potential impact of this temperature rise on livestock will include

1. Changes in yield and quality of feed and fodder produced : An increase of 2°C will produce negative impacts on pasture and livestock production in arid and semiarid regions and positive impacts in humid temperate regions. The length of growing season is also an important factor for forage quality and quantity because it determines the duration and periods of available forage. A decrease in forage quality can increase methane emissions per unit of gross energy consumed.

2. Reduced water availability will influence the livestock sector, which uses water for animal drinking, fodder crops and product processes. The livestock sector accounts for about 8% of global human water use and an increase in temperature may increase animal water consumption by a factor of two to three.
3. Variations in animal growth: Thermal livestock stress decreases feed intake and efficiency of feed conversion. Feed intake reduction also leads to a negative energy balance and reduced weight gain.
4. Disease occurrence: Animal health can be affected directly or indirectly by climate change, especially rising temperatures (Nardone *et al.,* 2010). The direct effects are related to the increase of temperature, which increases the potential for morbidity and death. The indirect effects are related to the impacts of climate change on microbial communities (pathogens or parasites), spreading of vector-borne diseases, food-borne diseases, host resistance and feed and water scarcity.
5. Changes in reproduction efficiency of both livestock sexes may be affected by heat stress. Oocyte growth and quality is affected, there is impairment of embryo development and pregnancy rate. Cow fertility may be compromised by increased energy deficits and heat stress. Heat stress has also been associated with lower sperm concentration and quality in bulls, pigs, and poultry.
6. Changes in production: High-producing dairy cows are more sensitive to heat stress. When metabolic heat production increases in conjunction with heat stress, milk production declines. Global warming may reduce body size, carcass weight and fat thickness in ruminants. Heat stress on birds will reduce body weight gain, feed intake and carcass weight. In pigs, protein and muscle calorie content will decline.
7. Experimental studies showed that milk yield of crossbred cows in India (e.g. Karan Fries, Karan Swiss and other Holstein and Jersey crosses) to be negatively correlated with temperature-humidity index. The average daily milk yield of the crossbred animals in the hot–humid eastern part of the country was significantly reduced by the rise in minimum temperature and not maximum temperature, as rise in minimum temperature crossed the critical temperature of comfort while the maximum temperature was already above the comfort zone. The rising temperature decreased the total dry matter intake and milk yield in Haryana cows. The productivity of Sahiwal cows also showed a decline due to increase in temperature and relative humidity buffaloes (Sirohi and Michaelowa, 2007).

Feed nutrient utilization and feed intake

Livestock have several nutrient requirements including energy, protein, minerals, and vitamins, which are dependent on the region and type of animal (Thornton et al., 2009). Failure to meet the dietary needs of cattle during heat stress affects metabolic and digestive functions (Mader, 2003). Sodium and potassium deficiency under heat stress may induce metabolic alkalosis in dairy cattle, increasing respiration rates (Chase, 2012).

Thermal stress

Thermal stress decreases feed intake and efficiency of feed conversion (McDowell, 1968), especially for livestock that are fed large amounts of high quality feeds (Haun, 1997). In the case of cattle, feed intake reduction leads to a negative energy balance and reduced weight gain (Lacetera *et al.,* 2003). Reduction of water intake may also decrease sweating and feed intake (Henry *et al.,* 2012).

Animal production

High-producing dairy cows are more sensitive to heat stress as they generate more metabolic heat than low-producing dairy cows. Metabolic heat production increases in conjunction with heat stress and milk production declines (Berman, 2005). Heat stress also affects ewe, goat, and buffalo milk production (Finocchiaro *et al.,* 2005).

Reproduction

Reproduction efficiency of both livestock sexes may be affected by heat stress. In cows and pigs, it affects oocyte growth and quality (Barati *et al.,* 2008), impairment of embryo development, and pregnancy rate (Nardone *et al.,* 2010). Cow fertility may be compromised by increased energy deficit and heat stress (King *et al.,* 2006). Heat stress has also been associated with lower sperm concentration and quality in bulls, pigs, and poultry (Kunavongkrita *et al*., 2005).

Livestock diseases

The effects of climate change on livestock diseases depend on the geographical region, land use type, disease characteristics, and animal susceptibility. Prolonged high temperature may affect metabolic rate (Webster, 1991), endocrine status (Johnson, 1980), oxidative status (Bernabucci *et al.,* 2002), glucose, protein and lipid metabolism, liver functionality [reduced cholesterol and albumin] (Bernabucci *et al*., 2006; Ronchi *et al.,* 1999), non-esterified fatty acids-NEFA (Ronchi *et al*., 1999), saliva production, and salivary HCO content. In addition, greater energy deficits affect cow fitness and longevity (King *et al*., 2006).

Animal health can be affected directly or indirectly by climate change, especially rising temperatures (Nardone *et al*., 2010). The direct effects are related to the increase of temperature, which increases the potential for morbidity and death. The indirect effects are related to the impacts of climate change on microbial communities (pathogens or parasites), spreading of vector-borne diseases, food-borne diseases, host resistance, and feed and water scarcity (Thornton *et al*., 2009). There is high probability that emergence of new diseases may act as a mixing vessel between human and livestock, facilitating combination of new genetic material and their transmissibility. This makes it difficult to estimate actual disease risk because of the dependence of diseases on animal exposure and interactions factors (Randolph, 2008).

Green House Gas Emissions

The primary livestock GHG emissions are CO_2, nitrous oxide (N_2O), and methane (CH_4), contributes the most to anthropogenic GHG emissions (44%), followed by N_2O (29%) and CO_2 (27%) [Gerber *et al*., 2013]. Globally livestock contribute 44% of anthropogenic CH, 53% of anthropogenic N_2O and 5% of anthropogenic CO_2 emissions.

The Global Livestock Environmental Assessment Model (GLEAM) developed by FAO-2017 and shows in CO_2 equivalents the GHG incidences that enteric fermentation and manure storage have across the main livestock species raised worldwide (Table 2).

Table 2: GHG incidence of enteric fermentation and manure storage by animal type

Animal types	Enteric methane (%)	Manure storage CH_4 (%)	Manure storage N_2O (%)	Total Gigatonnes CO_2 equivalents
Beef cattle	91	3	6	1.8 (45 %)
Dairy cattle	85	8	7	1 (26 %)
Buffaloes	91	2	7	0.5 (12 %)
Pigs	11	69	20	0.3 (7 %)
Sheep	93	3	4	0.2 (4.5 %)
Goat	93	4	3	0.2 (4 %)
Chicken	0	34	66	0.1 (1.5 %)

Source: FAO 2017

Enteric fermentation and manure account for 80% of 52 agricultural emission sources (Steinfeld *et al*., 2006). The enteric fermentation produced by ruminant livestock (e.g. cattle, sheep and goat) emits globally between 87 and 94 Tg of methane annually (IPCC, 2013). Mixed crop-livestock systems account for 64% of global enteric fermentation methane emissions; grazing systems account for 35, and industrial 1% (Steinfeld *et al.,* 2006). The countries that contribute the most methane emissions related to livestock production are India, China Brazil, and the United States (IPCC, 2013; Olivier and Janssens-Maenhout 2012). India, with the largest livestock population in the world, emitted 11.8 Tg of CH_4 in 2003, from which 91% derives from enteric fermentation and 9 % from manure management (Chhabra *et al.,* 2013).

Emissions by Species and Commodities

Most GHG emissions are from beef and dairy cattle, accounting for 65% of the total livestock GHG emissions. Pigs, poultry, buffaloes, and small ruminants contribute about 7 to 10%. If GHG emissions are estimated based on commodities, beef cattle contribute the most with 41% of the sector's emission, followed by dairy cattle (20%), swine (9%), buffalo (8%), poultry (8%), and small ruminant (6%). Enteric fermentation is the largest source of GHG emissions from cattle, buffalo, and small ruminants, comprising between 43% and 63% of the livestock sector emissions. However, for pigs and chickens the largest source of emissions is due to feed production (between 25% and 27%) which includes fertilizer production, machinery use, and feed transportation. Enteric fermentation from pigs is much lower than in ruminants because their digestive process does not produce as much methane as a by-product (Gerber *et al.,* 2013; Downing *et al,* 2017).

Manure

Livestock manure releases CH_4 and N_2O gas. The decomposition of the organic materials found in manure under anaerobic conditions releases methane (EPA, 1999). Estimated global methane emissions from manure decomposition of is 17.5 million tonnes of CH_4 per year. Pig manure comprises almost half of global manure-related methane emissions. At the country level, China has the highest global methane manure-related emissions, primarily due to pig manure. N_2O emissions from stored manure are equivalent to 10 million tonnes N per year. 1.7 million tonnes of manure soil N_2O are released per year. N_2O emissions from applied manure are 40% higher in mixed crop-livestock systems than the N_2O emissions from excreted manure deposited on pasture systems. Industrial production systems have 90% less N_2O emissions than mixed crop-livestock systems (Steinfeld *et al.,* 2006).

Climate smart agriculture

Sustainably increasing agricultural productivity and incomes, Adapting and building resilience to climate change

Climate-smart agriculture is an approach that helps to guide actions needed to transform and reorient agricultural systems to effectively support development and ensure food security in a changing climate. It aims to tackle three main objectives. *1.* Sustainably increasing agricultural productivity and incomes; *2.* Adapting and building resilience to climate change; and *3.* Reducing and/or removing greenhouse gas emissions.

Different elements which can be integrated in climate-smart agricultural approaches include: management of farms, crops, livestock, aquaculture and capture fisheries to manage resources better, produce more with less while increasing resilience.

In integrated production systems the products, by-products or services of one component of the system serve as a resource for the other production component. In these systems, the production components of the farm are mutually supportive and mutually dependent. An important aspect of integrated production is that the total production from the system is more important than the yield and/or efficiency of any individual production component. Integration of animals into farming systems not only provide milk and meat but also recycle their feed into manure that enhances the carbon sequestration.

Sustainable intensification of crop and livestock production systems with a strong market orientation is a key driver of increasing agricultural productivity and is crucial to addressing livelihoods, food security and the sustainable use of natural resources under changing climate conditions. Diversity appears to be critical in enhancing the resilience of agriculture in the face of climate change and the increasing intensity and frequency of extreme weather events.

Ecosystem and landscape management to conserve ecosystem services are key to increase at the same time improve resource efficiency and resilience. Climate smart agricultural options integrate traditional and innovative practices, technologies and services that are relevant for particular location to adopt climate change and variability.

In this context, agroforestry is recognized as an important component in climate-smart agriculture. Although agroforestry systems are not primarily designed for carbon sequestration, there are many recent studies that substantiate the evidence that agroforestry systems can play a major role in storing carbon in above ground and below ground biomass. Agroforestry systems are believed to have a higher potential to sequester C than pastures or field crops (Kirby and

Potvin, 2007). Trees and agroforestry systems provide a wide range of products and services that can substitute for each other and in the right circumstances, can be produced synergistically. Agroforestry can enhance livelihoods in rural communities by providing a variety of food, fodder and tree products, which increase food and nutritional security generate income and alleviate poverty. The restoration of degraded landscapes using agroforestry can increase the resilience of communities to shocks, including drought and food shortages and help mitigate climate change.

- About 30% of methane emissions at the global level are released from the ruminant activity. Rumen microbial utilization of carbohydrates in the gut of animals results in the production of volatile fatty acids, microbial protein, CO_2 and CH_4 with little hydrogen (H_2). Methane generation should be viewed as an energy sink where H_2 from all rumen microorganism's drains, allowing a greater total yield of ATPs.
- In India, 49.1% of enteric methane was contributed by cattle, 42.8% by buffaloes, 5.38% by goats and 2.59% by sheep. Importantly, during 1961–2010, the increase in methane emissions (70.6%) from livestock population of India is much greater than the increase in methane emissions from livestock population worldwide (54.3%). It is reported that by 2050, about 15.7% of enteric CH_4 emission at the global level will be contributed by the livestock population of India (Patra, 2014).
- CH_4 from enteric fermentation by ruminants is not only an important GHG associated with environmental problems, but it also represents a loss of feed energy (20–150 kJ/MJ) intakes (Singh *et al.*, 2005). Therefore, developing feeding strategies to minimize CH_4 emission is desirable in long-term mitigation of emission of GHG into the atmosphere and for short-term economic benefits.

Strategic Plans to Sustain Livestock Production under Changing Climate Scenario (Bhatta *et al.*, 2016)

Mitigation strategies to reduce livestock related GHGs

Generally, the methane mitigation strategies can be grouped under three broader headings viz., management, nutrition and advanced biotechnological strategies.

- Promoting research to reduce enteric methane through nutritional interventions – propionate enhancers, oil supplementation, ionophore supplementation etc.

- Reducing enteric methane emission using plant secondary metabolites
- Managing crop rotations that sequester carbon, conserve water and maintain soil fertility
- Improving feed resources that reduce greenhouse gas emissions
- Sustainable management of grazing in combination with fallowing and / or rehabilitation of degraded lands
- Exploring options for development of whole farm modeling for GHG emission
- Exploring the possibility of development of vaccines for methane reduction in ruminant livestock.

Strategies to reduce enteric methane emission

There are various strategies available to mitigate enteric methane emission. The strategies may be broadly grouped as management strategies, feeding strategies, rumen manipulation and advanced strategies.

The management strategies that could be adopted to control enteric methane emission are reducing the ruminant livestock population, breeding management and manure management. The most effective methane mitigating strategy will be to increase livestock productivity, which may permit reduction in livestock numbers to provide the same product output at a reduced enteric methane emission. Adopting a superior breeding program to enhance per animal productivity is one of the best options to reduce enteric methane emission as the production of enteric methane is greatly reduced on account of improvement in the productivity of livestock. Methane emitted from manures can be greatly reduced by minimizing the duration of storage of manure and giving lesser time for the "microbial fermentation" to take place. Anaerobic digesters are recommended as a mitigation strategy for methane, and are a source of renewable energy as well and provide sanitation opportunities for developing countries.

Feeding strategies are practical approaches to mitigate enteric methane emission and can be practiced with ease by farmers under field conditions. These may include pasture improvement, feed processing, increasing concentrate in ration and strategic supplementation. Feed processing can improve the feeding value of feeds by increasing its digestible energy content and / or by increasing feed intake. Therefore, an attempt to increase feed intake may reduce methane emission. These techniques are chopping and grinding of straws, alkali/ammonia treatment of straws and feed residues, urea molasses blocks. These processing techniques are reported to depress the methane emission from rumen by 10%.

In tropical countries, where ruminant livestock are grazed in wastelands or on very poor quality pastures or fed with poor quality agricultural waste, the digestibility of the diet is low and the nutritional requirement of the animals is not met. In this context protein supplementation in the diets increases the nutrient digestibility and significantly decreases methane production in the rumen (Mehra *et al.* 2006). Balanced diet with inclusion of protein supplements when fed to the lactating cows and buffaloes caused a marked reduction in the methane emission and an observable increase in both the production of milk and fat content in milk (Kannan and Garg, 2009).

Enteric methane emission from ruminant livestock can be reduced by various approaches that cause rumen manipulation viz., supplementation of bacteriocins, ionophores, fats, oils (Beauchemin *et al.* 2008), organic acids (Newbold *et al.*, 2005; Wallace *et al.* 2006), probiotics, prebiotics sulphate, halogenated methane analogues, nitroxy compounds (Morgavi *et al.*, 2010), fungal metabolites and secondary plant metabolites (Patra *et al.*, 2006; Wallace *et al.*, 2002; Calsamiglia *et al.*, 2007), microalgae, exogenous enzymes, etc.

Different livestock adaptation options to sustain production

- Developing and promoting drought – tolerant and early – maturing crop species.
- The adoption of improved animal breeds and grass / legume seed stock with increased resilience to projected climate conditions
- Adapt annual production cycle to better match feed production
- Adopting integrated disease surveillance response systems and emergency preparedness to prevent, mitigate, and respond to epidemic
- Strengthening meteorological services to provide timely weather and climate forecast /information early – warning systems.
- Promoting and strengthening aquaculture, poultry raising and the like as alternative livelihood options
- Developing and promoting guidelines for using herbal and alternative medicine
- Increasing agriculture extension activities for wide dissemination of knowledge about climate change impact
- Migration of livestock to better fodder production areas during drought
- The proactive management counter measures during heat waves (e.g. providing sprinklers or changing the housing pattern, etc.) or animal

nutrition strategies to reduce excessive heat loads are often expensive and beyond the means of small and marginal farmers who own most of the livestock in India. Moderate increase in temperature in high altitudes or winter months may decrease the maintenance requirement of animals (Sirohi and Michaelowa, 2007).

- Recent explosion of genotyping and resequencing data that are currently being used to develop genomically enhanced breeding approaches (Weller *et al.*, 2017). A large number of omics datasets (e.g. genomics, proteomics metabolomics, and transcriptomics) need to be used (Animal Agriculture 2019).
- Use of novel breeding schemes involving multiple *in vitro* rounds of genomic selection, gene editing, gamete production, and fertilization could reduce by orders of magnitude both the generation interval and the genetic lag between nucleus and commercial populations, based on embryonic stem cell technologies (Bogliotti *et al.*, 2018) and surrogate sire/dam technology (Park *et al.*, 2017; Taylor *et al.*, 2017). This lead to the development of a population of commercial animals that lack their own germline cells, but which carry transplanted gonial stem cells delivering the genetics from elite donor seed stock animals (Gottardo et al., 2018).
- How best to put into combination of phenotypes, environmental, and omics information in conjunction with gene editing and advanced reproductive technologies. Expectation is a 10-fold increase in the rate of genetic improvement in livestock, poultry, and aquaculture populations by 2030. There is also a need to develop objective selection criteria to enable the incorporation of important components of sustainability, like increased fertility, improved feed efficiency, functionality, and decreased susceptibility to disease into animal breeding programs (Animal Agriculture, 2019).
- Animal nutrition: Precision feeding, exploration of the microbiome and examination and implementation of novel feedstuffs. Precision feeding entails offering feed to each animal that is exactly tailored to the animal's needs.
- Use of human-inedible resources of slaughterhouse and food wastes, by products of biofuel production, leaf meals, seaweeds, and insect meals could replace human edible components of livestock diets. Plant breeding could also increase livestock production efficiency by (1) raising the feed crop yield per hectare (e.g., improved drought tolerance or nitrogen-use efficiency) and (2) improving the rate of feed conversion efficiency of vegetable calories into animal calories (e.g. altered digestibility or crop composition) such as reduced-lignin alfalfa (Van Eenennaam and Young, 2014). Electronic management of and feed conversion in ruminants grazing

on grasslands and range. Virtual fences are a new foray into this area and could improve grassland productivity and protect sensitive areas from overgrazing (Umstatter, 2011).

- A variety of technologies can be used to deal with the effects of short-terms heat waves, including shading or sprinkling to reduce excessive heat loads. Improved farmhouse micro-climate management through the use of thermal insulating construction materials and modern ventilation systems to protect livestock from extreme conditions and increase productivity.
- Improved pasture management by matching stocking rates to pasture production and integrating pasture improvement to increase feed value.
- Efforts are needed to develop suitable breeding program specific for climate change adaptation and mitigation. Projections suggest that further selection for breeds with effective thermoregulatory control will be needed. This calls for the inclusion of traits associated with thermal tolerance in breeding indices and more consideration of genotype-by-environment interactions (GxE) to identify animals most adapted to specific conditions.
- Selections for heat tolerance based on temperature – humidity index (THI) in genetic evaluation models are promising. Significance of including molecular markers and marker assisted selection in such breeding programs offers huge scope for developing in the most appropriate breed which can survive and reproduce normally in specific agro-ecological zone.

Summary

To meet future needs of an expanding human population, animal productivity will need to increase, however to control global warming and GHG emission intensity per unit of livestock product produced needs to decrease. As the numbers of farm animals reared for meat, egg, and dairy production increase, so do emissions from their production. By 2050, global farm animal production is expected to double from present levels. There is a great potential to reduce greenhouse gas emissions from livestock production. Although the reduction in GHG emissions from livestock industries are seen as high priorities, strategies for reducing emissions should not reduce the economic viability of enterprises if they are to find industry acceptability. Mitigating and preventing the environmental harms caused by this sector require immediate and substantial changes in regulation, production practices, and consumption patterns. To achieve this new research initiative at national and international levels on mitigation technologies is needed. Moreover, it is crucial to transfer successful technologies to farmers, monitor implementation and verify emissions from livestock post transfer. Intensify efforts for methane abatement from this sector as this would also be instrumental in increasing production of milk by reducing energy loss from the animals through methane emissions. Unlocking the potential for low carbon livestock requires concerted action by all stakeholders. The environmental impacts of animal agriculture require that governments, international organizations, producers, and consumers focus more attention on the role played by meat, egg, and dairy production. The impact of climate change can heighten the vulnerability of livestock systems, hence, ensuring good animal welfare, practicing climate smart agricultural practices especially related to livestock production and propagating resilient breeds suited to the local environment can to a certain extent tackle the problem.

International to individual levels: Enacting strict laws, acts, rules and regulations at the global, central and State levels to reduce the process of climate change by man-made activities. Cooperation of all can improve the chances of reduction of climate change in India. To follow simple steps like cleanliness of surroundings, proper management of daily garbage, reduce water wastage, reduce the use of plastics, judicious use of electricity, reuse of garbage, reduce the emission of harmful gases and discharge of chemicals from industry and last but not the lease, plant more trees (Shepali, 2019).

References

Alexandratos, N., Bruinsma, J., 2012. World agriculture towards 2030/2050: the 2012 revision. ESA Working paper No. 12–03. FAO, Rome.

Animal Agriculture. 2019. In: Sceice Breakthroughs to Advance Food and Agricultural Research by 2030. National Academy of Sciences. Washington D.C. pp. 57-82.

Aydinalp, C., Cresser, M.S., 2008. The effects of climate change on agriculture. Agric. Environ. Sci. 5, 672–676.

Barati, F., Agung, B., Wongsrikeao, P., Taniguchi, M., Nagai, T., Otoi, T., 2008. Meiotic competence and DNA damage of porcine oocytes exposed to an elevated temperature. Theriogenology 69, 767–772.

Beal, T., Massior, E., Arsenault, J.E., Smith, M.R., Hijmans, R.J.2017.Global trends indietary micronutrient supplies and estimated prevalence of inadequate intakes. Plos ONE. https://doi.org/10.1371/journal.pone.0175554

Bellarby, J., Tirado, R., Leip, A., Weiss, F., Lesschen, J.P., Smith, P., 2013. Livestock greenhouse gas emissions and mitigation potential in Europe. Glob. Change Biol. 19, 3–18.

Berman, A.J., 2005. Estimates of heat stress relief needs for Holstein dairy cows. J. Anim. Sci. 83, 1377–1384.

Bernabucci, U., Lacetera, N., Ronchi, B., Nardone, A., 2002. Markers of oxidative status in plasma and erythrocytes of transition dairy cows during hot season. J. Dairy Sci. 85, 2173–2179.

Bernabucci, U., Lacetera, N., Basirico, L., Ronchi, B., Morera, P., Seren, E., Nardone, A., 2006. Hot season and BCS affect leptin secretion of periparturient dairy cows. Dairy Sci. 89, 348–349.

Bhatta. R, V. Sejian and P. M. Malik, 2016. Livestock and climatic change: Contribution, impact and adaptation from India contest, Innovative designs, implements for global environment and entrepreneurial needs optimizing utilitarian sources, Indigenous, 94-110.

Bogliotti, Y.S., Wu, J., Vilarino, M., Okamura, T. et al., 2018. Efficient derivation of stable primed pluripotent embryonic stem cells from bovine blastocysts. Proc Natl Aca Sci USA, 115: (9): 2090-2095.

Canfield, D.E., Glazer, A.N., Falkowski, 2010. The evolution and future of Earth's nitrogen cycle. Science, 330 (6001): 192-196.

Carpenter, S.R., Caraco, N.F., Correll, D.L., Howarth, R.W., Sharpley, A.N. and Smith, V.H.1998. Nonpoint pollution of surface waters with phosphorus and nitrogen. Ecological Applications, 8(3): 559-568.

Chase, L.E., 2012. Climate change impacts on dairy cattle. Climate change and agriculture: Promoting practical and profitable responses.

Chhabra, A., Manjunath, K.R., Panigrahy, S., Parihar, J.S., 2013. Greenhouse gas emissions from Indian livestock. Climatic Change 117, 329–344.

Daghir, N. J., 2009. Nutritional Strategies to Reduce Heat stress in Broiler and Broiler Breeders. Lohaman Information. 44(1): 6.

DeMeester, S. L., T. G. Buchman and J. P. Cobb, 2001. The heat shock paradox: does NF-kappaB determine cell fate? FASEB J. 15(1): 270 - 274.

Downing, M.M.R., A. Pouyan Nejadhashemi, Timothy Harrigan, Sean A. Woznicki, 2017. Climate change and livestock: Impacts, adaptation, and mitigation, Climate Risk Management 16: 145–163.

FAO (Food and Agriculture Organization of the United Nations), 2009a. Global agriculture towards 2050. High Level Expert Forum Issues Paper. FAO, Rome.

Fathi, M. M., A. Galal , S. El-safty and M. Mahrous,2013. Naked neck and frizzle genes for improving chickens raised under high ambient temperature: I. Growth performance and egg production ,World's Poultry Science Journal , 69: 813-831.

Finocchiaro, R., van Kaam, J., Portolano, B., Misztal, I., 2005. Effect of heat stress on production of dairy sheep. J. Dairy Sci. 88, 1855–1864.

Gottardo, p., Gorjanc, G., Battagin, M., Gaynor, G.R. Jenko, J. et al., 2018. A strategy to exploit surrogate sire technology in animal breeding programs. Biorxiv, 199893. https://doi,org/10.1101/199893.

Gerber, P.J., Steinfeld, H., Henderson, B., Mottet, A., Opio, C., Dijkman, J., Falcucci, A., Tempio, G., 2013. Tackling Climate Change Through Livestock: A Global Assessment of Emissions and Mitigation Opportunities. FAO, Rome.

Haun, G.L., 1997. Dynamic responses of cattle to thermal heat loads. J. Anim. Sci. 77, 10–20.

Henry, B., Charmley, E., Eckard, R., Gaughan, J.B., Hegarty, R., 2012. Livestock production in a changing climate: adaptation and mitigation research in Australia. Crop Pasture Sci. 63, 191–202.

Hurst, P., Termine, P., Karl, M., 2005. Agricultural workers and their contribution to sustainable agriculture and rural development. FAO, Rome.

IPCC (Intergovermental Panel on Climate Change), 2007. Climate Change 2007: Synthesis Report. In: Pachauri, R.K., Reisinger, A. (Eds.), Contribution of Working Groups I, II and III to the Fourth assessment report of the Intergovernmental Panel on Climate Change. IPCC, Geneva, Switzerland, p. 104.

IPCC (Intergovermental Panel on Climate Change), 2014. Climate Change 2014: impacts, adaptation, and vulnerability. part A: global and sectoral aspects. In: Field, C.B., Barros, V.R., Dokken, D.J., Mach, K.J., Mastrandrea, M.D., Bilir, T.E., Chatterjee, M., Ebi, K.L., Estrada, Y.O., Genova, R.C., Girma, B., Kissel, E.S., Levy, A.N., MacCracken, S., Mastrandrea, P.R., White, L.L. (Eds.), Contribution of Working Group II to the Fifth Assessment Report of the Intergovernmental Panel on Climate Change. Cambridge University Press, Cambridge, United Kingdom and New York, NY, USA, p. 1132.

Jahejo, A. R., N. Rajput, N. M. Rajput, I. H. Leghari, R. R. Kaleri, R. A. Mangi, M. K. Sheikh and M. Z. Pirzado, 2016. Effects of Heat Stress on the Performance of Hubbard Broiler Chicken. Cells, Animal and Therapeutics, 2(1): 1 - 5.

James, A.e., Palmer, G.H. 2015. The role of animal source foods in improving nutritional in urban informal settlements. Identification of knowledge gaps and implementation barriers. Journal of Child Health and Nutrition, 4: 94-102.

King, J.M., Parsons, D.J., Turnpenny, J.R., Nyangaga, J., Bakari, P., Wathes, C.M., 2006. Modellig energy metabolism of Friesians in Kenya smallholdings shows how heat stress and energy deficit constrain milk yield and cow replacement rate. Anim. Sci. 82, 705–716.

Kunavongkrita, A., Suriyasomboonb, A., Lundeheimc, N., Learda, T.W., Einarsson, S., 2005. Management and sperm production of boars under differing environmental conditions. Theriogenology 63, 657–667.

Lacetera, N., Bernabucci, U., Ronchi, B., Nardone, A., 2003. Physiological and productive consequences of heat stress: The case of dairy ruminants. Proc. of the Symposium on Interaction between Climate and Animal Production: EAAP Technical Serie, pp. 7, 45–60.

Lin, H., K. Mertens, B. Kemps, T. Govaerts, B. De Ketelaere, J. De Baerdemaeker, E. Decuyperel and J. Buysel, 2004. New approach of testing the effect of heat stress on eggshell quality: mechanical and material properties of eggshell and membrane. Br Poult Sci. 45(4): 476 - 482.

Mader, T.L., 2003. Environmental stress in confined beef cattle. J. Anim. Sci. 81, 110–119.

May, J. D and B. D. Lott, 1992. Feed and water consumption patterns of broilers at high environmental temperature. Poult Sci. 71: 331-336.

McDowell, R.E., 1968. Climate versus man and his animals. Nature 218, 641–645.

Mosier, A., Kroeze, C., Nevison, C., Oenema, O., Seitzinger, S., Van Cleemput, O., 1998. Closing the global N2O budget: Nitrous oxide emissions through the agricultural nitrogen cycle: OEDC/IPCC/IEA phase II development of IPCC guideline for national greenhouse gas methodology. Nutr. Cycl. Agroecosyst. 52, 225–248.

Mosier, A., Wassmann, R., Verchot, L., King, J., Palm, C., 2004. Methane and nitrogen oxide fluxes in tropical agricultural soils: sources, sinks and mechanisms. Environ. Dev. Sustainability 6, 11–49.

Mottet, A., G.Tempio.2017. Global poultry production: Current state and future outlook and challenges. World's Poultry Science J., 73(2):245-256.

Nardone, A., Ronchi, B., Lacetera, N., Ranieri, M.S., Bernabucci, U., 2010. Effects of climate change on animal production and sustainability of livestock systems. Livest. Sci. 130, 57–69.

Olfati Ali, Ali Mojtahedin, Tayebeh Sadeghi, Mohsen Akbari and Felipe Martínez-Pastor, 2018. Comparison of growth performance and immune responses of broiler chicks reared under heat stress, cold stress and thermoneutral conditions. Spanish J Agri Res. 16(2): 505 - 512.

O'Mara, F.P., 2012. The role of grasslands in food security and climate change. Ann. Bot-London 110, 1263–1270.

Park, K.E., Kaucher, A.V., Powell, A., Waqas, M.S. et al. 2017. Generation of germline ablated male pigs by CRISPR/Cas9 editing of the NANOS2 gene. Scientific Reports, 7: 40176.

Rajkumar, U., A. Vinoth, M. Shanmugam, K. S. Rajaravindra, and S. V. Rama Rao, 2015. Effect of embryonic thermal exposure on heat shock proteins (Hsps) gene expression and serum T3 concentration in two broiler populations. Anim. Biotechnol. 26(4): 260 - 267.

Randolph, S.E., 2008. Dynamics of tick-borne disease systems: minor role of recent climate change. Rev. Sci. Technol. Oie. 27, 367–381.

Ronchi, B., Bernabucci, U., Lacetera, N., Verini Supplizi, A., Nardone, A., 1999. Distinct and common effects of heat stress and restricted feeding on metabolic status in Holstein heifers. Zootec. Nutr. Anim. 25, 71–80.

Rosegrant, M.W., Fernandez, M., Sinha, A., 2009. Looking into the future for agriculture and AKST. In: McIntyre, B.D., Herren, H.R., Wakhungu, J., Watson, R.T. (Eds.), International Assessment of Agricultural Knowledge, Science and Technology for Development (IAASTD). Agriculture at a crossroads, Island Press, Washington, DC, pp. 307–376.

Saif, Y. M., H. J. Barnes, J. R. Gilsson, A. M. Fadly, L. R. McDonald and D. E. Swayne, 2003. In: Disease of poultry. 11th Ed., lowa State Press, A Blackwell Publishing Co. Ames, lowa. p. 1056.

Sejian, V., J. B. Gaughan, R. Bhatta and S.M.K. Naqvi, 2016. Impact on climate change on livestock productivity. www.feedipedia.org.

Shepali, S. 2019. Climate change and India. Dream 2047, August, 2019. pp. 30-31.

Sivaramakrishnan. S., A.V. Omprakash, S. Ezhilvalavan, K.G. Tirumurugaan and A. Varun, 2017. Successful development of thermo tolerance in NANDANAM BROILER-3 chicken through thermal conditioning, Proceedings of Indian Poultry Science Association Conference 28-30th November 2017, pp- 95.

Sivaramakrishnan. S., A.V. Omprakash, S. Ezhilvalavan, K.G. Tirumurugaan and A. Varun, 2017. Effect of thermal conditioning on the production performance of different strains of chicken, Proceedings of Indian Poultry Science Association Conference 28-30th November 2017, pp-227.

Sivaramakrishnan. S., A.V.Omprakash, S. Ezhil Valavan, K.G.Thirumurugan and A.Varun, 2018. Effect of Thermal Conditioning on the Production Performance of different Strains of Chicken, International Journal of Livestock Research, 8 (2): 291-298.

Sivaramakrishnan. S., A.V. Omprakash, S. Ezhilvalavan, K.G. Tirumurugaan, G. Srinivasan and A. Varun, 2018. Evaluation of Heat Tolerance Potential by the Expression of HSP-70 mRNA in Different Chicken Strains through Thermal Conditioning, National conference on Native chicken production: opportunities for conservation, productivity enhancement and commercial exploitation in view of global warming held at Madras Veterinary College, Chennai on 19th & 20th December 2018.

Sivaramakrishnan. S., A.V. Omprakash, S. Ezhilvalavan, K.G. Tirumurugaan, G. Srinivasan and A. Varun, 2018. Evaluation of triiodothyronine and corticosterone hormone in early thermal conditioned chicken strains under the thermal conditioning and thermal challenge, National conference on Native chicken production: opportunities for conservation, productivity enhancement and commercial exploitation in view of global warming held at Madras Veterinary College, Chennai on 19th & 20th December 2018.

Smita, S., Michaelowa, A. 2007. Sufferer and cause: Indian livestock and climate change. Climatic Change, 85: 285-298.

Steinfeld, H., Gerber, P., Wassenaar, T., Castel, V., Rosales, M., Haan, C., 2006. Livestock's Long Shadow: Environmental Issues and Options. FAO, Rome.

Syafwan, S., R. P. Kwakkel and M. W. A. Verstegen, 2011. Heat stress and feeding strategies in meat-type chickens. Worlds Poult. Sci. J. 67: 653 - 673.

Taylor, L., Carlso, D.F., Nandi, S., Sherman, A., Fahrenkrug, S.C. McGrew, MJ. 2017. Efficient TALEN-mediated gene targeting of chicken primordial germ cells. Development, 144(5): 928-934.

Thomas, C.D., Cameron, A., Green, R.E., Bakkenes, M., Beaumont, L.J., Collingham, Y.C., Erasmus, B.F.N., de Siqueira, M.F., Grainger, A., Hannah, L., Hughes, L., Huntley, B., van Jaarsveld, A.S., Midgley, G.F., Miles, L., Ortega-Huerta, M.A., Peterson, A.T., Phillips, O.L., Williams, S.E., 2004. Extinction risk from climate change. Nature 427, 145–148.

Thornton, P.K., Van de Steeg, J., Notenbaert, A., Herrrero, M., 2009. The impacts of climate change on livestock and livestock systems in developing countries: A review of what we know and what we need to know. Agric. Syst. 101, 113–127.

Umstatter, C.2011, The evolution of virtual fences: A review. Computers and Electronics I Agriculture, 75(1): 10-22.

UN (United Nations), 2013. World population projected to reach 9.6 billion by 2050. United Nations Department of Economic and Social Affairs.

Van Eenennam, A.L., Young, A.E. 2014. Prevalence and impacts of genetically engineered feedstuffs on livestock productions. J Anim Sci., 92: 4255-4278.

Varun, A., 2019. Influence of heat shock proteins on the production and immunity parameters in native chicken germplasm, Ph.D., thesis submitted to the TANUVAS, Chennai (unpublished data).

Varun, A., A.V. Omprakash, K. Kumanan, S. Vairamuthu and N. Karthikeyan. Effect of thermal conditioning on physiological response of native chicken. The Indian Veterinary Journal, June 19, Accepted.

Varun, A., C. Yogesh, P.V. Sangeetha, S. Ezhil Valavan, M. Thangapandiyan, C. Pandian and A.V. Omprakash. Diagnosis of lymphoid leucosis in an organized aseel farm. The Indian Veterinary Journal, May 19, Accepted.

Webster, A.J., 1991. Metabolic responses of farm animals to high temperature. In: Ronchi, B., Nardone, A., Boyazoglu, J. (Eds.), Animal Husbandry in Warm Climates. EAAP Publication, pp. 15–22.

Weller, J.I., Ezra, E., Ron, M.2017. Invited review: A perspective on the future of genomic selection in dairy cattle. J Dairy Sci., 100: 8633-8644.

2

Climate Vulnerability and Resilient Agriculture in Odisha

Pravat Kumar Roul

Climate change is now a global phenomenon and its impact on livelihood health and wellbeing, and overall quality of life is not deniable. No country is free from the overall impacts of climate change, but poor people of developing countries have been disproportionately affected by the adverse effects. Vulnerability in the context of climate change is a function of sensitivity, exposure and adaptive capacities. Odisha is the 9th largest state by area and the 11th largest state by population in India. The state has an area of 155,707 km^2, which is 4.87 per cent of the total area of India, and a coastline of 480 km. In the eastern part of the state lies the coastal plain. It extends from the Subarnarekha river in the north to the Rushikulya river in the south. The State is broadly divided into four geographical regions, i.e. Northern Plateau, Central River Basins, Eastern Hills and Coastal Plains. The climate of the state is characterized by hot summer and cold winter in the interior parts. The state has historically been highly prone to climate change and multiple hazards – mainly cyclones, droughts and floods. Natural disasters devastate millions of lives and livelihoods in Odisha each year. More children and women suffer from the effects of natural disasters and this is predicted to worsen as storms floods and droughts become more severe and frequent because of climate change.

The natural calamities in Odisha during 1964 to 2019 are as follows:

Calamity	Year
Drought	1965, 1966, 1976, 1979, 1984, 1987, 1996, 1998, 2002, 2004, 2005, 2010, 2012, 2017, 2018
Flood	1969, 1970, 1971, 1973, 1975, 1977, 1985, 1990, 2001, 2003, 2007, 2008, 2009
Super Cyclone, ESCS (Fani)	1999, 2019
Cyclone and Flood	1967, 1968, 1971, 2013, 2014, 2018
Drought and Flood	1972, 1974, 1980, 1992, 2000, 2006, 2011, 2015
Drought, Flood and Cyclone	1982
Drought, Flood, Whirlwind & Tornado	1981
Hailstorm, Tornado and Whirlwind	1978

Extreme temperature, frequent and intensive flood, cyclone and other natural disasters due to climate change has become acute and expecting to be severe in future. Besides, Odisha is mainly rainfall dependent as its irrigation network does not cover the entire state. The agriculture sector is vulnerable to the vagaries of climate-induced weather changes. Food security is also threatened in different parts of Odisha due to climate change induced disasters. Rise in temperature and sea level has made agriculture vulnerable as the gushing sea water combined with erratic rain often destroys the crops. Sea water is more often gushing into the agricultural land, filling it with saline water, which is directly affecting farmers and slowly weakening the productivity of the state. Agriculture across the coast of Odisha is now facing a serious climate emergency. The climatic variations could further multiply the vulnerability of poor by adversely affecting their health and livelihoods and impeding the development of the state.

Climate Change and Resilience

The development of resilient agricultural systems is an important aspect as many communities greatly depend on the provisioning ecosystem services of such systems (food, fodder, and fuel) for their livelihoods (Altieri, 1999). Many agricultures based economies have limited alternate livelihood strategies (Tilman et al., 2002), and small family farms have little capital to invest in expensive adaptation strategies, which increases the vulnerability of rural, agricultural communities to a changing environment. The challenge for the researchers and extension personnel is to develop resilient agricultural systems using rational, affordable strategies such that ecosystem functions and services can be maintained and livelihoods can be protected. Thus, a resilient agro-ecosystem will continue to provide a vital service such as food production if challenged by severe drought or by a large reduction in rainfall. In agricultural systems, diversification may provide the link between stress and resilience because adversity of organisms is required for ecosystems to function and provide services (Heal, 2000). Vandermeer et al. (1998) elucidated the main issues linking the role of diversity in agro-ecosystems to functional capacity and resilience.

It is evident that climate change in Odisha has the potential to tremendously aggravate water stress, food security and health system. In the same time resilient agriculture including crop diversification could be an effective adaptation option under this situation as it protects natural biodiversity, strengthening the ability of the agro-ecosystem to respond to these stresses, minimizing environmental pollution, reducing the risk of total crop failure, reducing incidence of insect pests, diseases and weed problems and secure food supply opportunities and also providing producers with alternative means of generating income. It provides opportunities for food security and enables farmers to grow surplus products for sale at market and thus obtain increased income to meet other needs related to household well-being. Thus the knowledge on selecting suitable enterprise, cropping systems, crops and varieties is of importance for climate change adaptation.

Crop Diversification for Resilience

Diversification in agriculture essentially refers to a shift from one crop/ variety/ cropping system to another or from one enterprise to another. The first one is Crop diversification, while the second one is Enterprise diversification. The crop diversification may be a necessity due to one or more of the following reasons.

- To overcome aberrant weather
- To use the available resources
- To improve the soil fertility status
- To reduce the menace of pest
- To generate employment and income
- To meet the market demand

It is an important instrument for economic growth. The diversified cropping systems have not only minimized the climatic and market risks, but have also resulted in providing diverse and nutritious food at the household level. Efforts, therefore, need to be made to explore fully the potential and prospects of crop diversification to forge congruence of enhanced productivity, profitability and sustainability.

Resilience, as defined by Walker et al., (2006) described it as the capacity of a system to experience shocks while retaining essentially the same function, structure, feedbacks, and therefore identity. The resilience of a system is also related to the magnitude of shock that the system can absorb and remain within a given state, the degree to which the system is capable of self-organization, and the degree to which the system can build capacity for learning and adaptation (Folke et al., 2002). Crop diversification can improve resilience in a variety of ways: by the ability to suppress pest outbreaks and dampen pathogen transmission, which may worsen under future climate scenarios, as well as by buffering crop production from the effects of greater climate variability and extreme events. Such benefits point toward the obvious value of adopting crop diversification to improve resilience, yet adoption has been slow.

Adaptation is increasingly regarded as a key component of the response to climate change, and occurs through adjustments to reduce vulnerability or enhance resilience in response to observed or expected changes in climate and associated extreme weather effects (Brooks and Adger, 2005). During the last several years, there have been a number of definitions for adaptation to climate change. Smithers and Smit (1997) defined adaptation as "changes in a system in response to some force or perturbation, in our case related to climate". The adoption of crop diversification has been found to be among some of the most common adaptation measures adopted by farmers in response to climate change (Verchot et al., 2007; Altieri and Koohafkan, 2008; Kissinger et al., 2013).

Major outcomes of effective crop diversification

- Knowledge generation
- Improved farming systems
- Improved dietary variation
- Use of new technologies and practices (e.g. soil and water conservation, pest and disease control)
- New sources of energy (fuel)
- New plant varieties developed and released
- New plants (for food, fuel, fodder and forage) and animal breeds introduced
- Income generation
- Increased labour productivity
- Farmer groups formed, thus farmer empowerment
- Women's empowerment and gender equity

- Conservation of plant genetic resources improved (including collection and protection) Exchange of plant genetic resources enhanced
- New linkages established among actors in the value chain
- New niche market opportunities developed
- Increased investment of the private sector in value added to crops and development of the value chain
- Increased public awareness of the importance of biodiversity and the benefits of diversification
- New policies formulated to support crop diversification
- Better coordination and cooperation among countries in the region

Constraints in Crop Diversification

The major problems and constraints in crop diversification are primarily due to the following reasons with varied degrees of influence:

- Over 117 m/ha (63 percent) of the cropped area in the country is completely dependent on rainfall.
- Sub-optimal and over-use of resources like land and water resources, causing negative impact on the environment and sustainability of agriculture.
- Inadequate supply of seeds and plants of improved cultivars.
- Fragmentation of land holding less favoring modernization and mechanization of agriculture.
- Poor basic infrastructure like rural roads, power, transport, communications etc.
- Inadequate post-harvest technologies and inadequate infrastructure for post-harvest handling of perishable horticultural produce.
- Very weak agro-based industry.
- Weak research - extension - farmer linkages.
- Inadequately trained human resources together with persistent and large scale illiteracy amongst farmers.
- Host of diseases and pests affecting most crop plants.
- Poor database for horticultural crops.
- Decreased investments in the agricultural sector over the years.

India is already under pressure from climate stresses which increase vulnerability to further climate change and reduce adaptive capacity. The adverse effects of climate change have a particularly devastating effect on agriculture, which is the mainstay of Indian economies. This has affected food production with its resultant effect on widespread poverty. Some farming communities have developed traditional agricultural adaptation strategies including enterprise and crop diversification options to cope with climate variability and extreme events. Experience with these strategies needs to be shared among communities with validation by researchers. An important aspect in this direction is lack of general awareness knowledge, expertise and data on climate change issues. Actions to address these gaps include: training programmes for local government officials, dedicated research activities and academic programmes; and the initiation of specific institutional frameworks for climate change. Furthermore, improving and strengthening human capital, through education, outreach, and extension services, improves decision-making capacity at every level and increases the collective capacity to adapt.

References

Altieri, M. A. and Koohafkan, P., 2008. Enduring farms: Climate change, smallholders and traditional farming communities. Third World Network, Penang, Malaysia.

Brooks, N. and Adger, W. N., 2005. Assessing and enhancing adaptive capacity. In: Lim B, Spanger- Siegfried E, editors. Adaptation Policy Frameworks for Climate Change: Developing Strategies, Policies and Measures. UNDP-GEF.Cambridge University Press Cambridge, UK. pp. 165–181.

Kissinger, G., Lee, D., Orindi, V. A., Narasimhan, P., Kinguyu, S. M.and Sova, C., 2013. Planning Climate Adaptation in Agriculture. Meta-Synthesis of National Adaptation Plans in West and East Africa and South Asia. CCAFS Report No.10. CGIAR Research Program on Climate Change, Agriculture and Food Security, Copenhagen, Denmark.

Smithers, J. and Smit, B., 1997.Human adaptation to climatic variability and change. Global Environmental Change,7(2):129– 46.

Tilman, D., Cassman, K. G., Matson, P. A., Naylor, R., and Polasky, S., 2002. Agricultural sustainability and intensive production practices. Nature, 418: 671–677.

Vandermeer, J., van Noordwijk, M., Anderson, J., Ong, C. and Perfecto, I., 1998. Global change and multi-species agroecosystems: Concepts and issues. Agriculture, Ecosystems and Environment, 67: 1–22

Verchot, L. V, Noordwijk, M., Kandji, S., Tomich, T., Ong, C. and Albrecht, A., 2007. Climate change: Linking adaptation and mitigation through agroforestry. Mitigation and adaptation strategies for global change, 12(5):901–18.

Walker, B., Gunderson, L., Kinzig, A., Folke, C., Carpenter, S. and Schultz, L., 2006.A handful of heuristics and some propositions for understanding resilience in social-ecological systems.Ecology and Society, 11(1)

3

Climate Smart Livestock Management System: Livestock Advisories to Negate the Impacts of Climate Change

B S Rath

Climate change may manifest itself as rapid changes in climate in the short term or subtler changes over decades. Generally, climate change is associated with an increasing global temperature. Various climate model projections suggest that by the year 2100, mean global temperature may be 1.1–6.4 °C warmer than in 2010. The difficulty facing livestock is weather extremes, e.g. intense heat waves, floods and droughts. In addition to production losses, extreme events also result in livestock death (Gaughan and Cawsell-Smith, 2015). Animals can adapt to hot climates, however the response mechanisms that are helpful for survival may be detrimental to performance. In this article we make an attempt to project the adverse impact of climate change on livestock production. There is a two-way relationship between livestock production and environmental health.

Impact of climate change on livestocks

The potential impacts on livestock include changes in production and quality of feed crop and forage, water availability, animal growth and milk production, diseases, reproduction, and biodiversity. These impacts are primarily due to an increase in temperature and atmospheric carbon dioxide (CO_2) concentration, precipitation variation, and a combination of these factors. Temperature affects most of the critical factors for livestock production, such as water availability, animal production, reproduction and health. Forage quantity and quality are affected by a combination of increases in temperature, CO_2 and precipitation variation. Livestock diseases are mainly affected by an increase in temperature and precipitation variation.

Quantity and quality of feeds

Quantity and quality of feed will be affected mainly due to an increase in atmospheric CO_2 levels and temperature. The effects of climate change on quantity and quality of feeds are dependent on location, livestock system, and species. Some of the impacts on feed crops and forage are:

- The effects of increasing concentration of CO_2 on herbage growth will be positive due to inducing partial closure of stomata, reducing transpiration, and improving some plants' water-use efficiency; with greater effect on C3 than C4 species
- Changes in temperature and CO_2 levels will affect the composition of pastures by altering the species competition dynamics due to changes in optimal growth rates. Primary productivity in pastures may be increased due to changes in species composition if temperature, precipitation, and concurrent nitrogen deposition increase.
- Quality of feed crops and forage may be affected by increased temperatures and dry conditions due to variations in concentrations of water-soluble carbohydrates and nitrogen. Temperature increases may increase lignin and cell wall components in plants, which reduce digestibility and degradation rates, leading to a decrease in nutrient availability for livestock.
- Extreme climate events such as flood, may affect form and structure of roots, change leaf growth rate, and decrease total yield.

Water availability

Water availability issues will also influence the livestock sector, which uses water for animal drinking, feed crops, and product processes.

- The livestock sector accounts for about 8% of global human water use and an increase in temperature may increase animal water consumption by a factor of two to three . To address this issue, there is a need to produce crops and raise animals in livestock systems that demand less water or in locations with water abundance.
- As sea level rises, more saltwater will be ingressed into coastal freshwater aquifers. Water salination could affect animal metabolism, fertility, and digestion.
- Chemical contaminants and heavy metals could impair cardiovascular, excretory, skeletal, nervous and respiratory systems, and impair hygienic quality of production.

Livestock diseases

The effects of climate change on livestock diseases depend on the geographical region, land use type, disease characteristics, and animal susceptibility. Animal health can be affected directly or indirectly by climate change, especially rising temperatures. The direct effects are related to the increase of temperature, which increases the potential for morbidity and death. The indirect effects are related to the impacts of climate change on microbial communities (pathogens or parasites), spreading of vector-borne diseases, food-borne diseases, host resistance, and feed and water scarcity. Climate change may induce shifts in disease spreading, outbreaks of severe disease, or even introduce new diseases, which may affect livestock that are not usually exposed to these type of diseases.

Heat stress

All animals have a thermal comfort zone, which is a range of ambient environmental temperatures that are beneficial to physiological functions. Heat stress on livestock is dependent on temperature, humidity, species, genetic potential, life stage, and nutritional status.

- When temperature increases more than the upper critical temperature of the range (varies by species type), the animals begin to suffer heat stress.
- Animals have developed a phenotypic response to a single source of stress such as heat called acclimation. Acclimation results in reduced feed intake, increased water intake, and altered physiological functions such as reproductive and productive efficiency and a change in respiration rate.
- Warm and humid conditions cause heat stress, which affects behavior and metabolic variations on livestock or even mortality.

Biodiversity

Climate change may eliminate 15% to 37% of all species in the world. Temperature increases have affected species reproduction, migration, mortality, and distribution. The livestock species that had the highest percentages of risk of breed elimination were chicken (33% of breeds), pigs (18% of breeds) and cattle (16% of breeds). However, the breeds at risk depend on the region. Developing regions had between 7% and 10% of mammalian species at risk (not restricted to livestock), but between 60% and 70% of mammalian species are classified as of unknown risk. Conversely, in developed regions where the livestock industry is very specialized and based on a small number of breeds, the mammalian species at risk were between 20% and 28% states that this biodiversity loss is mainly because of the practices used in livestock production that emphasize yield and economic returns and marginalization of traditional production systems where other considerations are also important (such as ability to withstand extremes).

Adaptation strategy

Adaptation and mitigation are two complementary strategies for responding to climate change. Adaptation is the process of adjustment to actual or expected climate and its effects, in order to either lessen or avoid harm or exploit beneficial opportunities. Mitigation is the process of reducing emissions or enhancing sinks of greenhouse gases, so as to limit future climate change. Both adaptation and mitigation can reduce and manage the risks of climate change impacts. Adaptation measures involve production and management system modifications, breeding strategies, institutional and policy changes science and technology advances, and changing farmers' perception and adaptive capacity.

Breeding strategies

It is obvious that the production oriented breeding strategies has to be chaned to stress resistance strategy. Changes in breeding strategies can help animals increase their tolerance to heat stress and diseases and improve their reproduction and growth development. Therefore, the challenge is in increasing livestock production while maintaining the valuable adaptations offered by breeding strategies, all of which will require additional research.

Adjustment in production practices

The integrated farming system approach should be developed by diversification intensification and/or integration of pasture management, livestock and crop production together. It will reduce risk and enhance resilience to climate change by modifying microclimate, enhancing soil productivity and availability of feed and fodder.

Livestock management systems: To reduce the direct impact the focus must be given on

(i) designing high quality, low-cost natural shade and improving water availability to reduce heat stress from increased temperature,

(ii) changes in livestock/herd composition (selection of large animals rather than small);

(iii) reduction of livestock numbers. a lower number of more productive animals leads to more efficient production and lower GHG emissions from livestock production

(iv) developing infrastructure to harvest and store rainwater, such as tanks connected to the roofs of houses and small surface and underground dams.

Restoration of Community Grazing Resources

Community grazing resources (CPR) in most parts of the country has either drastically been reduced or has almost disappeared (Misra *et al.,* 2014). Restoration is most important not only for providing regulated grazing but also for protecting valuable resource. One option is restoration of such lands to village panchyat for their improvement and proper management through silvi pastoral development. CPR may be subjected to one of the grazing systems namely (i) Continuous grazing, (ii) Deferred grazing, (iii) Rotational grazing and (iv) Deferred rotational grazing for a given site at a specific period as per requirement.

Alternative feed resources

Inclusion of alternate feed resources in the animals. diet could be a useful strategy to minimize the nutritional stress during lean months. It has been demonstrated that feeding of Cactus, and *azollae* to livestock reduced water requirement and increased nutrients digestibility without affecting health of the calves in arid or semi-arid regions.

Capacity building for livestock keepers

There is a need to improve the capacity of livestock producers and herders to understand and deal with climate change increasing their awareness of global changes. In addition, training in agro-ecological technologies and practices for the production and conservation of fodder improves the supply of animal feed and reduces malnutrition and mortality in herds.

Livestock advisories for different weather extremes

In the changing scenario of climate advisory for livestock production and management systems are barely necessary to address the adverse effect of climate change. Some of the advisories for different extreme weather events rae discussed here

Drought: Drought is a condition where there is shortage of precipitation for a sufficient longer period of time. The waters in rivers, streams and underground maybe lower than average causing hydrological imbalance. Drought results in shortage of feeds and fodders and the animal may remain in stress hence reducing their productivity.

Management during drought

- Early warnings help better preparations for drought mitigation strategies
- Early plans should involve veterinary health care institutions, water resources and disaster assistance to expand their services in times of need
- Provision for additional water supply in times of water shortages
- Explore the use of conventional feed and fodder resources and encourage supply of molasses to cattle feed plants. Dry fodder reserves, urea molasses licks, bricks made of fodder urea and molasses etc. can be part of the stock pile.
- Implementing measures to stabilize fodder resources by using seed reserves and planting alternative drought resistant fodder crops
- Prevention of disease outbreak through better health management protocol

Flood - Floods are one of the most common natural disasters causing extensive damage to property, livestock, crops and human life. But animals are natural swimmers; therefore can escape drowning if they are not tied or caged. In an event of flood, the environment, drinking water and rivers become contaminated. The fear of outbreak of infectious diseases like tetanus, dysentery, hepatitis and food poisoning etc. becomes prominent with poor management.

Management during flood

- Evacuate the animals rapidly to higher ground and check for injuries to be attended by a veterinarian
- Prevent the outflow of manure pits into rivers or even drinking water
- All local ponds and canals should be inspected for any obstruction
- Ensure that the animals are vaccinated for all infectious diseases
- The animals should be brought to safer places if the forecast of a disaster is beforehand. In flooded areas where drainage is slow, can be used for duck rearing and fish farming

Cyclones: Meteorologically, cyclone can be predicted with some accuracy. Hence the loss can be avoided through better preparedness of an event.

Management during flood

- Cyclone shelters can be made to house animals away from the cyclonic area
- Animals should be shifter to higher grounds
- Stocking of concentrates and medicines

- Vaccination of animals
- Make provisions for early disposal of carcasses

Future perspectives

Responding to the challenges of global warming necessitates a paradigm shift in the practice of agriculture and in the role of livestock within farming systems. Science and technology are lacking in thematic issues, including those related to climatic adaptation, dissemination of new understandings in rangeland ecology (matching stocking rates with pasture production, adjusting herd and water point management to altered seasonal and spatial patterns of forage production, managing diet quality, more effective use of silage, pasture seeding and rotation, fire management to control woody thickening and using more suitable livestock breeds or species), and a holistic understanding of pastoral management (migratory pastoralist activities and a wide range of bio-security activities to monitor and manage the spread of pests, weeds, and diseases). Integrating grain crops with pasture plants and livestock could result in a more diversified system that will be more resilient to higher temperatures, elevated carbon dioxide levels, uncertain precipitation changes, and other dramatic effects resulting from the global climate change. The key thematic issues for effectively managing environment stress and livestock production include:

- Development of early warning system;
- Research to understand interactions among multiple stressors;
- Development of simulation models;
- Development of strategies to improve water-use efficiency and conservation for diversified production system;
- Exploitation of genetic potential of native breeds; and
- Research on development of suitable breeding programmes and nutritional interventions.

The integration of new technologies into the research and technology transfer systems potentially offers many opportunities to further the development of climate change adaptation strategies. Epigenetic regulation of gene expression and thermal imprinting of the genome could also be an efficient method to improve thermal tolerance.

4

Climate Change and Its Impact on Extreme Weather Events, Livestock and Agriculture

S C Sahu

Introduction

Climate change has been realised during 1840s when indisputable evidence of former ice ages was obtained. The climate has altered sufficiently in many parts of the world even within the last few decades which will affect the possibilities for agriculture and settlement. Reliable weather records have been kept only during the last hundred years or so, but *proxy* indicators of past conditions from tree rings, pollen in bog and lake sediments, ice core records of physical and chemical parameters, and ocean foraminifera in sediments provide a wealth of paleoclimatic data. The standard interval adopted by the World Meteorological Organization for climatic statistics is thirty years:1901–30, 1931–60, for example. However, for historical records and proxy indicators of climate, longer, arbitrary time intervals may be necessary to calculate average values. Tree rings and ice cores can give seasonal/annual records, while peat bog and ocean sediments may provide records with only 100- to 1,000-year time resolution. Hence, short-term changes and the true rates of change may, or may not, be identifiable.

The present climatic state is usually described in terms of an average value (arithmetic mean, or the median value in a frequency distribution), a measure of variation about the mean (the standard deviation, or interquartile range), the extreme values, and often the shape of the frequency distribution. A change in climate can occur in several different ways, for example, there may be a shift in the mean level, or a gradual trend in the mean values. The variability may be periodic, quasi-periodic or non-periodic, or alternatively it may show a progressive trend . First, it is important to determine whether such changes are real or whether they are an artefact of changes in instrumentation, observational practices, station location, or the surroundings of the instrumental site, or due to errors in the transcribed data. Even when changes are real, it may be difficult to ascribe them to unique causes because of the complexity of the climate system. Natural variability operates over a wide range of time scales, and superimposed on these natural variations in climate are the effects of human activities.

Cause of global warming

One of the most important topics in Meteorology and Climatology is global warming. Global warming is the theory that states when greenhouse gases are added to the earth's atmosphere, the result will be for increased average global temperature. The factors responsible for climate change are called Climate forcing. The variations in solar radiation, deviations in the earth's orbit, mountain building & continental drift and changes in greenhouse gases concentrations are included in climate forcing. There are a variety of climate change feed backs such as positive feedback and negative feedback which can amplify or diminish the initial forcing. Some parts of the climate system, such as the oceans and ice caps, respond slowly in reaction to climate forcing because of their large mass. Therefore, the climate system can take centuries or longer to fully respond to new external forcing. The main greenhouse gas that is of concern is Carbon Dioxide. Moreover, among greenhouse gases, methane is up to 20 times more effective in trapping heat in the atmosphere than carbon dioxide. It is created from a variety of natural and human-influenced sources which include landfills, natural gas, and petroleum sources. The scientists calculated the animals would have collectively produced more than 520 million tonnes of methane a year-more than all today's modern sources put together. Dinosaurs may have partly responsible for causing climate change as they emitted the potent global warming gas, methane. It is believed that the giant animals spent 150 years emitting methane. In fact, large plant-eating sauropods would have been the main culprits due to the huge amounts of greens they consumed and after breaking down in the animal's stomach it would have produced thousands of litres of the greenhouse gas compared with a modern cow which only produces 200 litres of methane daily, according to the scientists. The climate change was so catastrophic that it caused the dinosaurs eventual demise. There are some reasons which increase the global temperature mainly due to deforestation and forest fires. The replacing of trees and vegetation with pavement also contributes to warmer surface temperature. The total increase in terms of tons of carbon was 564 million, which according to the Chicago Sun Times article is higher than the individual emissions of all but three countries across the globe. Those three countries are China, India and the U.S. According to the U.S. Department of Energy, from 2009 to 2010 there was a whopping 6% increase in carbon released into the atmosphere globally. It is the highest percentage of increase since available record of 1751. It is due to burning of fossil fuels. The state level carbon emission in India are given below:

Table 1: State-level CO_2 Emissions: 2000

Sl.No.	State	Aggregate	Per Capita
		Metric tons of carbon	
1	Jammu and Kashmir	696.5	0.07
2	HP	659.1	0.11
3	Punjab	10845.7	0.45
4	Haryana	5460.5	0.26
5	Uttar Pradesh	44268.3	0.27
6	Rajasthan	8929.3	0.16
7	Delhi	6033.8	0.44
8	Bihar	9012	0.11
9	Orissa	16172.3	0.44
10	West Bengal	23363.7	0.29
11	Assam	1097	0.04
12	Gujarat	18461.5	037
13	Maharashtra	35595.4	0.37
14	Goa	652.2	0.44
15	Madhya Pradesh	39729.4	0.66
16	Andhra Pradesh	30126	0.40
17	Karnataka	9059.6	0.17
18	Kerala	3034.2	0.10
19	Tamil Nadu	17584.9	0.28
20	Others	43712.6	0.62

Effects of climate change or global warming

The average state of the climate system is controlled by a combination of *forcings* external to the system (solar variability, astronomical effects, tectonic processes and volcanic eruptions), internal radiative forcings (atmospheric composition, cloud cover), anthropogenically induced changes (in atmospheric composition, surface land cover) and *feedback effects* (such as changes in atmospheric water vapour content or cloudiness caused by global temperature changes).It is useful to try and assess the magnitude of such effects, globally and regionally, and the time scales over which they operate.

The impacts of global warming on climate will be significant. As temperature warms, more moisture can be evaporated into the air. Semi-arid areas are becoming deserts because of increase in temperature which in turn evaporates critical moisture that is needed for the land. Dry climates are becoming drier and wet climates are becoming wetter. With more moisture evaporating into the air, where it does rain, the rain is more intense. This creates extreme flooding in areas the moisture advects into and rains out. Climate change may cause the collapse of Indus valley civilization around 4000 years ago. The study, combining the latest archaeological data along with state-of-the-art geoscience technologies, suggested that decline in monsoon rains led to weakened river dynamics and played a critical role both in the development and the fall of the Harappan culture which relied on river floods to fuel their agricultural surpluses.

Furthermore, air and water pollution can be directly attributed to the livestock sector and these are the largest contributor to global water pollution. The livestock sector is also one of the leading drivers of global deforestation and is linked to 75 percent of historic deforestation in the Brazilian Amazon rainforest. Nearly a third of biodiversity loss to date has been linked to animal agriculture. Further amplifying water and air pollution, global livestock produce seven to nine times more sewage than humans most of which is left untreated. They also discharge pesticides, antibiotics and heavy metals into water systems. Concentrated animal farming operations present additional public health risks to nearby communities as viral diseases may spread from sick livestock to humans and the increased use of antibiotics encourages antibiotic resistance. Irresponsible manure management from high-volume facilities risks aerosolizing faecal matter that may reach nearby homes and cause respiratory problems. Livestock waste can also pass through the soil to groundwater, which may then contaminate nearby streams and rivers with nitrates and pathogens.

Many weeds, pests and fungi thrive under warmer temperatures, wetter climates and increased CO_2 levels. The ranges and distribution of weeds and pests are likely to increase with climate change. This could cause new problems for farmers› crops previously unexposed to these species. More extreme temperature and precipitation can prevent crops from growing. Extreme events, especially floods and droughts, can harm crops and reduce yields. Heat stress affects animals both directly and indirectly. Over time, heat stress can increase vulnerability to disease, reduce fertility, and reduce milk production. Drought may threaten pasture and feed supplies. Drought reduces the amount of quality forage available to grazing livestock. Some areas could experience longer, more intense droughts, resulting from higher summer temperatures and reduced precipitation. For animals that rely on grain, changes in crop production due to drought could also become a problem. Climate change may increase the prevalence of parasites and diseases that affect livestock. The earlier onset of spring and warmer winters could allow some parasites and pathogens to survive more easily. In areas with increased rainfall, moisture-reliant pathogens could thrive.

The global average temperature for May,2012 is the 2nd hottest ever since 1880 as compared to the hot experienced in May,2010. US National Climatic Data Centre has said that such a hot May was never recorded in the Northern Hemisphere. Recent analysis shows that 2016 is the hottest year and so, every year seems to be hottest as compared to the years already passed. It is believed that these extreme weather phenomena are mainly due to the rise in global temperatures. But the misery of rising heat is being felt worldwide with normal weather systems in disarray. If large areas of the western Himalayas in Uttarakhand have suffered raging forest fires, so has the US-more than 8 lakh hectares have been engulfed in flames. The March-May period for the US has been the hottest ever. Brazil is in the midst of its worst drought in five decades with more than 1,000 towns suffering. Heavy downpours and freak hail has hit China and flash floods have ravaged crops in Ethiopia.

Due to climate change, natural disasters are in increasing trend. Disaster is referred as any sudden, unexpected or extraordinary misfortune regardless of whether it occurs to an individual, a family or other small groups, a community, a region, a nation or the entire world. Natural disasters affect countries large and small, rich or poor, whatever their political persuasion. The toll exacted by natural calamities each year drains the human and economic resources of every nation and stands as one of the formidable barriers of national, regional and world development. It is well known fact that natural disasters are acts of God but losses due to them are acts of man. Lot of human suffering and misery from a large number of natural disasters can be mitigated by taking timely actions, preventing mechanisms and undertaking capital works of long and medium terms. The social and economic losses of disasters are very high and often immeasurable. The poor and the marginalized like the small and landless farmers and the agricultural labourers are the most sufferers by this. Due to climate change, frequency and intensity of natural disasters are now very much concerned now a days by the general public. We have to analyse in the context of Orissa for different type of natural disasters.

Moreover, the devastating toll of natural disasters is rising each year, in spite of our increased understanding of hazards and how to mitigate them. Five factors contribute to this growing potential for disaster.

1. Rise of world population and concentrate in large cities and other areas of high risk, such as flood plains, coasts, landslide-prone slopes and seismic zones which tend to magnify the impact of a hazard when it occurs.

2. Increased capital development has accompanied population growth. This translates into more structures of greater value- that is more to lose when a hazard strikes.
3. We human beings actually contribute to the severity of many hazards by altering our natural environment.
4. Our increasing vulnerability is the great number of hazardous dwellings and public buildings in use today. Though our knowledge to design structures to resist hazards has improved but most new structures do not incorporate this knowledge. That is, the number of older structures remains high and the job of raising them all to an ideal standard of safety is daunting.
5. Disasters today are shared events with repercussions far outside the immediately stricken area. This rising global risk will fall with greatest impact on the developing nations where the overwhelming majority of hazard-related fatalities already occur and economic losses- proportionally speaking are highest.

Hazards in Odisha

Odisha is located in the Eastern coast roughly between Lat. $17^0 49'$ N and $22^0 36'$ N and Long. $81^0 36'$ E and $87^0 18'$ E. Odisha is one of the most disastrous prone states in India as well as also in the world. In view of its tropical location and the long coast line, Odisha is vulnerable to major natural hazards like Cyclones, floods, heat waves, cold waves and droughts. Besides these, some parts of Odisha also fall under seismic zone III. Due to increase in extreme events, now time has come to predict the extreme weather just before 3 hours of it occurrence for saving the life of human beings and animals. Keeping in view that, IMD has developed now casting system in India including Odisha. To achieve the goal, 55 Doppler Weather Radar (DWR) networks are being established in India and four such systems at Balasore, Gopalpur, Sambalpur and Paradeep will be installed in Odisha during next 2 to 3 years.

Due to global warming, there is decline in monsoon rain in Odisha especially for last 4 to 5 years and rainfall is also not uniform in the whole monsoon season. So, both drought and flood are experienced in the monsoon season as in case of 2010 and 2011 monsoon rain. Untimely excess rain is caused due to climate change and which is responsible for the damage of crops in Odisha.

India with its long coastline is vulnerable to the impacts of tropical cyclones that develop in North Indian Ocean (Bay of Bengal and the Arabian Sea). These systems are classified as depressions, deep depressions, cyclonic storms, severe cyclonic storms, very severe cyclonic storms and super cyclonic storm.

A cyclonic storm is a rotational low pressure system which forms over warm ocean surface in tropics and is a vast violent whirl 150 to 800 Km across 10 to 17 Km high, spiralling around a centre in anticlockwise direction in Northern Hemisphere and progressing along the surface of the sea at a rate of 300 to 500 Km a day. In a low pressure system when the central pressure falls by 5 to 6 hPa from the surrounding and wind speed reaches 34 Knots we call it a cyclonic storm. Tropical cyclones are macro scale systems with meso-scale impact. Since the extent of the core of the cyclone hardly exceeds 100Km, the major impacts are generally localized.

The storm surges, strong winds and flooding due to torrential rains associated with cyclones are the major causes of destruction and damage. Cyclone Warning Centre (CWC) Bhubaneswar is responsible for the Odisha Coast in the East Coast of Bay of Bengal. People of Odisha especially coastal districts of Jagatsinghpur and Kendrapara have not yet forgotten the tragedy of 1999 super cyclone which took the life of about 10000 people and 3,70,297 cattleheads besides damaging crops about 16,17,000 hectares.

Odisha is a flood prone state and predominantly a flat deltaic and river irrigated land. The whole Odisha is criss-crossed by seven rivers the Mahanadi, the Subarnarekha, the Budhabalang, the Brahmani, the Baitarani, the Bansadhara and the Rusikulya with their tributaries which ripped the state. Rainfall is abundant from June to September and occasionally tropical cyclone strikes the coastal area, bringing the torrential rainfall. Odisha faces flood every year during monsoon. Unprecedented floods in 2008 and 2011 in Mahanadi will tell about the devastating phenomena in Odisha.While normal monsoon floods are beneficial for the state, severe one damages agriculture, infrastructure and also human lives. The short range forecasts are useful in deciding an early release of water in a phased and regulated manner so as to avoid sudden floods and consequent misery. There should be regular dialogue between the reservoir manager on the one hand and the operational meteorologists and hydrologists on the other, before appropriate management decisions are taken.

Heat wave is a recurring phenomena in Odisha. The last couple of decades have seen several of the hottest years on record. Global warming plus local warming are contributors. As places urbinize, they heat up (urban heat island). Temperature sensors in an urbanizing area will warm more than they otherwise would. Add to this global warming and it is nosurprise that the hottest years in history have occurred recently. As compared to other years, heat wave in 2012 is extreme in Odisha. Both coastal and interior districts experienced unprecedented continuously 32 days heat wave in Odisha which affected the normal life of human beings as well as domestic animals. All time record temperature 46.7 degree Celsius is observed at Bhubaneswar on 5th June, 2012.

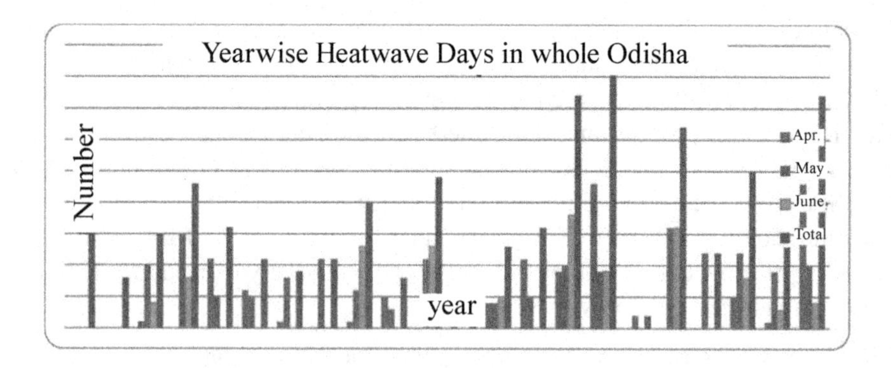

Conclusion

In view of the global environmental changes, it is likely that the frequency and impact of disasters will increase the world over. The population pressure is causing degradation of environment by interrupting the water flow, hydrological cycles, causing either landslides, floods, soil erosion etc. As a welfare state, the Government will have to take the lead in disaster prevention and reduction and mitigating their impact, enhancing the awareness of the coping mechanisms among the people and to prevent loss of lives and property.

The public awareness will have to be also created through the NGOs apart from the local administration. It should be the combined effort of the Government at Centre, the State, the District and the Panchayats. NGOs and people are to pool their resources, capability and put in their best efforts to face the situation and to mitigate the losses.

References

Annual and seasonal climatic variations over the northern hemisphere and Europe during the last century, Goossens.C & Berger.A,Annales Geophysicae,1986,4,B,4,385-400.

Damage Potential of Tropical Cyclone, Govt. of India, India Meteorological Department, October, 2002, pp1-29. IMD Data

The yearly averaged air temperature in Belgrade from 1888 to 1987, Microslava. U, Atmosfera (1990), 3, pp. 305-313.

5

Climate Change and Extreme Weather Events

Sandeep Pattnaik

Introduction

The Intergovernmental Panel on Climate Change (IPCC) assessment reports (IPCC, 2007) suggests that consequently occurring climate extremes are likely to increase the frequency and intensity in the future. In the Indian context, events such as landfall of rare of rarest most intense pre-monsoon Tropical cyclone Fani (2019), back to back landfall of very intense tropical cyclones such as Phalini (2013) and Hudbud (2014), Cyclone Komen (2015), Extreme rainfall events of Kerala (2018 and 2019), Heavy rainfall events of over Northeastern states of India (2019), Cloud burst events of Uttarakhand (2013), Rainfall over Himachal Pradesh (2019), Mumbai Rain fall (2015) and extreme heat wave conditions of 1998, 2015 and many similar type of extreme events are not usual. However, at present there is no unanimity among scientists to exclusively and directly link such extreme weather events to climate change, however based on number of studies and observation data sets it is agreed that the impact of climate change is evident (beyond the natural variability) in many part of the globe including India. In 2019 monsoon season at least 1351 deaths happened to floods, heavy rainfall and landslides compared to 1550 for year 2018. In this manuscript, recent research findings related to extreme weather events in the context India are discussed in the following section-2, need for high resolution and long term data sets are emphasized in section-3, followed by recommendations as part of future preparedness, that will not only enable us to address these issues in a robust scientific manner but also provide us strong resilient society to face these extreme weather events in an optimized and effective manner to minimize the loss of lives, properties and socio-economic conditions of citizens.

Goswami *et al.* (2006) has shown an increasing trend of extreme rainfall events over central India. Though later studies (Ghosh *et al.*, 2009, 2012) demonstrated disparity and in homogeneity in the trends of extreme rainfall over Indian region, however number of research results are coherent in their view that the frequency as well as intensity of severe weather systems are increasing over India. Using high resolution (0.25 × 0.25) gridded data sets Vinnarasi and Dhayna (2016) has shown that there is a significant increase in the maximum intensity of rainfall and spatial heterogeneity has been noted for past half century within their analysis of 113 years. They have also noted a shift in the frequency of distribution of extreme rainfall events during the monsoon period and significant negative trends in wet spell durations and positive trends in dry spell durations. Dash *et al.* (2011) has found that, the frequency of short spell rain events with heavy intensity have increased and long spell rain events with moderate to low intensities have decreased over India. Further, they found that short spell heavy to moderate intensity rain events have increased its contribution to the seasonal rainfall. Recently, Roxy *et al.* (2017) has shown threefold increase in extreme rainfall event over Indian region and this has been attributed to surge in monsoon westerlies, facilitating the enhanced moisture incursion to the Indian region. Baisya and Pattnaik (2019) investigated the Kerala heavy rainfall event of year 2018 and noted that multiple factors such as western-ghat orography, anomalous transportation of moisture through monsoon flow, traversing of moisture flux convergence towers from a monsoon depression over eastern India to the Kerala region and positive quasi bi-weekly/Intra seasonal oscillations are working in tandem for making this kind of severe heavy rainfall situation.

Using gridded temperature data sets, Ross *et al.* (2018) has shown that there is a consistent warming pattern is found over northwestern and southern India and a pattern of cooling was noted over the northeastern and southwestward regions of India. The pattern of warming has been attributed to the global warming scenario and cooling due to large scale present of haze and aerosol over the region. Mukherjee and Mishra (2018) recently found that there is a significant increase in 1 and 3-day concurrent hot day and hot night events over India. They have used model simulations from climate of 20th century plus detection and attribution and coupled model intercomparsion project 5 (CMIP5). Further, they have suggested that restricting global mean temperature below 1.5 from the pre-industrial level may substantially reduce the risk of 1 to 3 day concurrent hot day and hot night events over India. Mishra *et al.* (2017) has shown that if the global mean temperature is limited to 2.0 °C above pre-industrial era, there are chances that frequency of severe heat waves might increase by 30 times the current climate scenarios over Indian region. Using 30 years data sets, Keval and Pattnaik (2019), has noted that the daily maximum, minimum and mean temperature has increased at the rate of 0.006 °C, 0.012 °C and 0.017 °C per year, respectively in the districts of western Odisha and neighboring Chhattisgarh. Further, they noted that, over these regions, frequency and intensity of warm nights have increased, whereas frequency and intensity of cold nights have decreased over the years. Dash and Mamgain (2011) using IMD data sets have shown that there is a significant decrease in frequency of occurrence of cold nights in the winter months in India and in its homogeneous regions in north except over western Himalaya. Additionally, they have found that southern regions of India shows a drastic decrease in frequency of cold nights and increase in warm nights over the interior peninsula regions with highest region of warming over west coast of India. Ali and Mishra (2017) proposed that a strong positive relationship found between dew point and tropospheric temperature with rainfall extremes over urban locations in India. They have also suggested that surface air temperature might not be ideal parameters to examine the extreme rainfall cases over urban locations.

The world has witnessed the major heat wave events in the recent past (Rahmstorf and Coumou 2011, Sun *et al.* 2014). Especially over densely populated urban areas across the globe including India, where most of the human population is vulnerable to the severe impacts of these heat wave events (Mishra *et al.* 2015, Matthews *et al.* 2017, Dash and Mamgain 2011, Rohini *et al.* 2016). The tragic and severe heat wave events of India have occurred during 1998 and 2015 causing massive loss of lives. For 1998 (2015), the total number of reported deaths was 1300 (>2500), out of which 650 (35) deaths happened in Odisha (Rohini *et al.* 2016). In addition, the heat wave events are is one of the major factors for the climate migration of population around the globe particularly over the rural areas (Burrows *et al.* 2016). Ross *et al.* (2018) has noted that the decadal mean temperature during 1 April to 31 May over India has increased by 1°C in the 2010s compared to the 1950s in addition, annual mean temperature trend over western Odisha and Chhattisgarh region is 1.5 °C/100 year over the period for 1901-2016. Pai *et al.* (2004, 2013) showed that during the decade (1991-2000) that there is a noticeable increase in the frequency high-frequency temperature extreme events (i.e. heat waves). Further, number of these studies also indicated that there is an increase in the frequency and intensity of severe heat waves over Odisha. Srivastava *et al.* (2014) have found that the annual frequency of hot days has an increasing trend (5.1 days/decade) over Odisha, and the daily maximum temperatures also increase significantly over Odisha. Analysis reveals that the mean temperature of the Odisha state has increased by ~0.3 °C during the past three decades (1981-2010) with the most accelerated warming (~0.9 °C) occurring during the recent decade (2001 to 2010).

For extreme rainfall case of Mumbai (2005), Bhaskar Rao and Ratna (2010) has shown that National Centre for Atmospheric Research (NCAR) MM5 model with analysis nudging for 12 hour could able to reproduce best simulation results up to 55 cm of rainfall in 24 hours at correct location of Mumbai. They have noted that intense monsoon westerlies at the lower level with dry air incursion at the middle level suppressed convection and leading to enhanced potential instability at lower levels causing this cloud burst situation. Madala *et al.* (2014) simulated few thunderstorm events over Gadanki using WRF model with an aim to understand the impact of PBL parameterization on these events. They noted that Mellor-Yamada-Janjic (MYJ), scheme produced the best thunderstorm and MYJ-Grell-Devenyi ensemble scheme (GD) combination captured the vertical extent of the convective updraft but the event has lag and lead of one and half hour than observed. Halder *et al.* (2015) discussed about role of better ice nucleation process and their impact on realistic representation of severe thunderstorm over Indian region. They have suggested that improvement in cloud ice generation has positive impact on cloud microphysics and associated dynamics through latent heat release and leading to improvement in thunderstorm prediction. Sisodiya *et al.* (2019) has demonstrated that using high resolution improved land assimilation system it is possible to improve the location specific severe thunderstorm event over the state of Odisha. Rajeevan *et al.* (2010) has undertaken a study to assess the impact of cloud microphysics on thunderstorm and concluded that there are major disparities among simulations due to different cloud microphysics parameterization. All the four microphysics schemes underestimated strength and vertical extend of updraft and lacks in replicating the downdrafts of the storm. They have concluded with strong emphasis on carrying out intense observational campaign targeting to understand cloud microphysical and land surface processes over Indian region.

Rajesh *et al.* (2016) has demonstrated that high resolution model with improvement in land state conditions over Indian region can better simulate the intense rainfall event to understand these high impact weather systems such as Uttarakhand event of 2013. It is found that accurate representation of land state with special reference to soil moisture and temperature, leading to unravel the channel of continuous supply of moisture incursion from Arabian Sea to the Uttarakhand region aggravating the intensity of the event. Dimri *et al.* (2017) in a review article identified that, combination of factors such as large scale and small scale meteorological forcings, orographic influence and geomorphological conditions are responsible for cloud burst events over Himalayan region. Majority of these events occurred at the elevation range of 1000 m to 2500 m within the valley folds of the southern rim of the Indian Himalayas. Murugavel *et al.* (2014), suggested that convective available potential energy might be the only factor responsible for pre-monsoon lightning events. They found that prevailing meteorological conditions along with regional orography playing a major role in inducing lightning activity over India. Pawar *et al.* (2017) found that in pre-monsoon thunderstorm inverted polarity charge structure of the storm has a good correlation with the dew point depression. In addition, they have noted that the high aerosol concentration with adequate ice nuclei plays a dominant role in inverted polarity charge structure in thunderclouds.

High Resolution Observations and Long Term Data Sets

High resolution spatio-temporal observation data is one of the major challenges for the weather and climate researchers for accurate understanding of the physical processes, validation of model results and to provide an authenticate representation of climate change scenario over the Indian region. Therefore, reliable long term observation data is key for the progress of weather and climate science in India. Rajeevan *et al.* (2006) has developed a gridded daily rainfall data at 1° × 1° horizontal resolution taking surface observations from 1803 stations across the country. These data is considered as one of the benchmark data sets for rainfall observations over India and extensively used by the research community and operational agencies across the globe for various purpose. This effort has been augmented by Pai *et al.* (2013, 2014) by bringing out high resolution daily rainfall data at a spatial resolution of 0.25° × 0.25° over Indian main land. They have considered about 6955 rain gauge station for developing this data sets. This data set is now extremely popular and authenticated by several researchers in terms of its representativeness and accuracy.

Mitra *et al.* (2009, 2013) had developed a moderate resolution (1° × 1°) daily rainfall data sets through merging of satellite rainfall dataset [e.g., Tropical Rainfall Measuring Mission (TRMM), Global Precipitation Measurement (GPM] along with estimates from IMD rain gauge. This data is immensely popular among research community across India and the globe due to additional information of local gauge rainfall data and rainfall estimates are available over the oceanic regions. In this segment, the high resolution outgoing longwave radiation (OLR) data developed at 0.25° × 0.25° resolution by Mahakur *et al.* (2013) has potential to provide information about cloud types causing rainfall and facilitate in unravelling convective cloud organization from mesoscale to synoptic scale over the Indian region. Apart from these specific data sets, plethora of different atmospheric and ocean data sets were obtained from observation campaign, satellites and remote sensing instruments are readily available in IMD, Meteorological and Oceanographic Satellite Data Archival

Centre (MOSDAC), Indian Space Research Organization (ISRO), IITM Pune and Indian National Centre for Ocean Information Services (INCOIS) for researchers for carrying out research and development activities.

Indian research community has taken lead in initiating several targeted observational campaigns with specific scientific questions relevant to the country in recent years. To name a few, Cloud-Aerosol Interactions and Precipitation Enhancement Experiment (CAPIEEX), Interaction of Convective Organization and Monsoon Precipitation, Atmosphere, Surface and Sea (INCOMPASS), South West Asian Aerosol - Monsoon Interactions (SWAAMI), Forecast Demonstration Project on Cyclone (Raj *et al.*, 2010), Pre-monsoon SAARC Severe Thunderstorm observation and regional modelling (STORM) project (Ray *et al.*, 2015), Continental Tropical Convergence Zone (CTCZ) (DST, 2008). These massive observation campaigns under the national monsoon mission program will surely prove to be a milestone in achieving desired objectives of the outstanding scientific challenges in weather and climate sciences in coming years through enriching scientific community with rare and valuable observation and modelling data sets. Each of these observation campaign has been oriented towards specific scientific questions, region/duration of observations, deployment of observation equipment and modelling framework. These efforts are multi institutional and multidisciplinary in nature, hence requires massive and sustained co-ordination in terms of science, planning, logistics, administration and timely execution. Successful completion of these observation campaigns, has highlighted the point that India has developed indigenous capabilities and authorities to take up these emerging challenges in future with an aim to understand and unravel finer details of many intriguing science questions related to monsoon and other important weather systems over the region. This vital information in terms of observations can be subsequently used in terms of data assimilation, accurate representation of physical processes, validations of the models with aim to improve their forecast skills at different spatio-temporal resolution for diverse weather and climate systems over the Indian region.

Prepare for Future

Here, few challenging gap areas are identified (Pattnaik 2019), which we should focus in an unified manner to prepare ourselves well for emerging challenges in weather and climate extremes in the country.

- Defining the objective criteria of extreme weather events based on meteorological, hydrological and socio-economic long term observation data sets at different scale i.e. local, regional, global. This is essential component to objectively quantify the linkage.

- Understanding the exact attributes (quantification) for occurrence of extreme weather events and establishing the scientific causes for the same. Though identifying the exact causes is quite difficult keeping in view the chaotic nature of the weather and climate physics and dynamics, however, efforts must be put to move forward in this direction.
- In the climate change scenario, there is an increase in weather and climate extreme events over India. Therefore, more emphasize should be given to develop better early warning systems for societal preparedness and put intense efforts in terms of observation, modelling and validation for accurate identification and understanding of key physical processes over vulnerable regions responsible for causing these extreme events.
- Policy makers should regular growth and developments of citizens particularly leaving in hilly areas as these area are more vulnerable to hydro-meteo-geological extreme events particularly foot-hills of Himalayas, Western Ghats and northern eastern hilly districts.
- The thrust must be on developing and adapting state of art tools, models and emerging techniques such as artificial intelligence, machine learning, augmented and virtual reality and other digital framework not only to generate forecast but also to customized and disseminate region specific information for different kind of weather systems over the targeted area.
- Multi-disciplinary and focused research is need of the hour to understand, quantify, characterize and predict these location extreme weather events and associated local and regional impacts such livelihood of the citizens in different sections i.e. agriculture, fishery, energy, transport, disaster management, policy planning etc.
- Keeping our country's need, major grand challenges to address extreme weather and climate events must be identified & feasible roadmap to achieve those objectives in short term and long term time scale should be prepared.
- More high density observation facilities must be established in the country in terms of test beds, observatories for validation and long term unbiased time series data. Further, more research is required to explore cost effective unmanned aerial vehicle (UAV) procedure to assess the value of obtaining observations for intense weather systems in remote areas over land as well as ocean in order to improve models.

- A fresh and innovative approach should be adapted by the operational and research agencies for data collection, archival and dissemination to provide a seamless, thriving and vibrant research ecosystem to the academicians, scientists and students in country.
- Sustainable framework needs to be built for engaging high quality skilled manpower in weather and climate sciences to maintain the momentum of improving our capabilities to find solutions to the outstanding problems (including extreme events) and develop better models with reasonable forecast accuracy with adequate lead time for the benefits of citizens.
- It is essential to adapt a culture where academician, researcher, forecaster relevant agencies and other stake holders should work coherently and continuously as a team to not only review the existing weather and climate services in the country but also to come out with innovative strategies to effective address the evolving challenges.
- An interface must develop to continuous formulate methods to incorporate the observational understanding to the modelling framework.
- Rigorous multidimensional framework must be established for major and emerging cities of the country through inclusive approach considering all stake holders to develop long term effectiveness approach to develop a resilient urban society to adopt to extreme weather events.
- Apart from thrust areas discussed in the article few other areas needs attention such as regional climate variability, drought, cold wave, lightning & thunderstorm, cloud burst and land slide, coastal erosion, sea level rise and coastal inundation, acute air pollution, hail and dust storms may be emphasized.

Sustained efforts and discipline practices in these directions will enable our scientists and researchers to meet grand challenges of our society and prepare themselves well to adapt to the growing demands in weather and climate services of India in the 21st century.

References

Ali, H. and Mishra, V., 2017, "Contrasting response of rainfall extremes to increase in surface air and dew point temperatures at urban locations in India", Scientific Reports, 7, 1228.

Bhaskar Rao, Dodla Venkata and Ratna, S. B., 2010, "Mesoscale characteristics and prediction of an unusual extreme heavy precipitation event over India using a high resolution mesoscale model", Atmospheric Research, 95, 255-269.

Burrows K and Kinney P 2016 Exploring the climate change, migration and conflict nexus; Int. J. Environ. Res. Public Health 13(4) 443.

Dash, S. K. and Mamgain, Ashu, 2011, "Changes in the frequency of different categories of temperature extremes in India", Journal of Appl. Met. and Clim., 50, 1842-1858.

Dimri, A. P., Chevuturi, A., Niyogi, D., Thayyen, R. J., Ray, K., Tripathi, S. N., Pandey, A. K. and Mohanty, U.C., 2017, "Cloudbursts in Indian Himalayas: A review", Earth-Science Reviews, 168, 1-23.

DST, 2008, "Continental Tropical Convergence Zone (CTCZ) Programme: Science Plan", Department of Science and Technology, New Delhi, India, p167. http://odis.incois.gov.in/images/ctczimages/docs/science_plan.pdf.

Ghosh, S., Das, Debasish, Kao, Shih-Chieh and Ganguly, Auroop R., 2012, "Lack of uniform trends but increasing spatial variability in observed Indian rainfall extremes", Nature Climate Change, 2, 86-91, doi: 10.1038/nclimate1327.

Ghosh, S., Luniya, Vishal and Gupta, Anant, 2009, "Trend analysis of Indian summer monsoon rainfall at different spatial scales",Atmos. Sci. Let., 10, 285-290.

Goswami, B. N., Venugopal, V., Sengupta, D., Madhusoodanan, M. S. and Xavier, P. K., 2006, "Increasing trend of extreme rain events over India in a warming environment", Science, 314, 1442-1445.

Haldar, M., Hazra, Anupam, Mukhopadhyay, P. and Singh, Devendraa, 2015, "Effect of the better representation of the cloud ice-nucleation in WRF microphysics schemes: A case study of a severe storm in India", Atmospheric Research, 154, 155-174.

Keval M and Pattnaik. S, 2019, "Spatio temporal patterns of surface temperature over western Odisha and eastern Chattisgarh", SN App. Sci. 1: 991. https://doi.org/10.1007/s42452-019-0986-2

Madala, S., Satyanarayana, A. N. V. and Rao, T. Narayana, 2014, "Performance evaluation of PBL and cumulus parameterization schemes of WRFAR Wmodel in simulating severe thunderstorm events over Gadanki MST radar facility-Case study", Atmospheric Research, 139, 1-17.

Mahakur, M., Prabhu, A., Sharma, A. K., Rao, V. R., Senroy, S., Singh, Randhir and Goswami, B. N., 2013, "A high-resolution outgoing longwave radiation dataset from Kalpana-1 satellite during 2004-2012", Current Science, 105, 8, 1124-1133.

Matthews T K, Wilby R L and Murphy C 2017 Communicating the deadly consequences of global warming for human heat stress; Proc. Natl. Acad. Sci. 114(15) 3861-3866.

Mishra V, Ganguly A R, Nijssen B and Lettenmaier D P 2015 Changes in observed climate extremes in global urban areas; Environ. Res. Lett. 10(2) 024005.

Mitra, A. K., Bohra, A. K., Rajeevan, M. N. and Krishnamurti, T. N., 2009, "Daily Indian precipitation analyses formed from a merged of rain-gauge with TRMM TMPA satellite derived rainfall estimates", Journal of the Meteorological Society of Japan, 87A, 265-279.

Mitra, A. K., Momin, I. M., Rajagopal, E. N., Basu, S., M. N., Rajeevan, M. N. and Krishnamurti, T. N., 2013, "Gridded daily Indian monsoon rainfall for 14 seasons: Merged TRMM and IMD gauge analyzed values", J. Earth Syst. Sci., 122, 5, 1173-1182.

Mukherjee, S. and Mishra, Vimal, 2018, "A six-fold rise in concurrent day and night-time heat waves in India under 2 °C warming", Scientific Reports, 8:16922, DOI:10.1038/s41598-018- 35348-w.

Murugavel, P., Pawar, S. D. and Gopalakrishan, V., 2014, "Climatology of lightning over Indian region and its relationship with convective available potential energy", International Journal of Climatology, 34, 11, 3179-3187.

Pai D S, Nair S A and Ramanathan A N 2013 Long term climatology and trends of heat waves over India during the recent 50 years (1961–2010); Mausam 64 585-604.

Pai D S, Thapliyal V and Kokate P D 2004 Decadal variation in the heat and cold waves over India during 1971-2000; Mausam 55 281-292.

Pattnaik Sandeep (2019): Weather Forecasting in India: Recent developments, Mausam, 70, 3 (July 2019), 453-464

Pawar, S. D., Gopalakrishnan, V., Murugavel, P., Veremey, N. E. and Sinkevich, A. A., 2017, "Possible role of aerosols in the charge structure of isolated thunderstorms", Atmospheric Research, 183, DOI:10.1016/j.atmosres.2016.09.016, 331-340

Raj, Y. E. A., Balachandran, S., Ramanathan, R. M. A. N., Geetha, B. Ramesh, K., Selvam, N., Guhan, M. V. and Rajanbabu D., 2010, "Forecast Demonstration Project (cyclone)-2010", IMD Report, http://www.imdchennai.gov.in/wxSummFinal.pdf

Rajeevan, M., Kesarkar, A., Thampi, S. B., Rao, T. N., Radhakrishna, B. and Rajasekhar, M., 2010, "Sensitivity of WRF cloud microphysics to simulations of a severe thunderstorm event over Southeast India", Ann. Geophys., 28, 603-619.

Rajeevan, M., Bhate, J., Kale, J. D. and Lal, B., 2006, "High resolution daily gridded rainfall data for the Indian region: Analysis of break and active monsoon spells", Curr. Sci., 91, 3, 296-30

Rajesh, P. V., Pattnaik, S., Rai, D., Osuri, K. K., Mohanty, U. C. and Tripathy, S., 2016, "Role of land state in a high resolution mesoscale model for simulating the Uttarakhand heavy rainfall event over India", Journal of Earth System Science, 125, 3, 475-498, doi:10.1007/s12040-016-0678-x.

Rahmstorf S and Coumou D 2011 Increase of extreme events in a warming world; Proceedings of the National Academy of Sciences 108(44) 17905-17909.

Ray, K. Bandopadhyay. B. K., Sen. B., Sharma. P., Warsi A. H., Mohapatra. M., Yadav B. P., Debnath. G. C., Stella. S., Das. S., Duraisamy. M., Rajeev. V.K., Barapatre, V., Paul. S., Shukla. P, Madan. R., Goyal. S., Das. A. K., Bhan. S. C., Chakravarthy. K. and Rathore. L.S. 2015, "Pre-monsoon thunderstorms 2015", No. ESSO/IMD/SAARC STORM PROJECT-2015/(01)(2015)/4.

Rohini P, Rajeevan M and Srivastava A K 2016 On the variability and increasing trends of heat waves over India. Sci. Rep. 6 26153.

Ross, R., Krishnamurti, T. N., Pattnaik, S. and Pai, D. S., 2018, "Decadal surface temperature trends in India based on a new high-resolution data set", Scientific Reports, 8, 7452, doi:10.1038 /s41598-018-25347-2.

Roxy, M. K., Ghosh, S., Pathak, A., Athulya, R., Terray, P., Murtugudde, R., Mujumdar, M. and Rajeevan, M., 2017, "A threefold rise in widespread extreme rain events over central India", Nat. Commun., 8, Article number: 708, 1-11, DOI: 10.1038/s41467-017-00744-9.

Sisodiya A, Pattnaik. S, Baisya H, Bhat. G S and Turner A.G., 2019, "Simulation of location specific severe thunderstorm events using high resolution land assimilation", Dynamics of Atmospheres and Oceans, 87, https://doi.org/10.1016/j.dynatmoce.2019.101098

Srivastava A K, Singh G P, Singh O P and Choudhary U K 2015 Recent variability and trends in temperatures over India; Vayumandal 40 161-181

Sun Y, Zhang X, Zwiers F W, Song L, Wan H, Hu T, Yin H and Ren G 2014 Rapid increase in the risk of extreme summer heat in Eastern China; Nat. Clim. Change. 4(12) 1082.

Vinnarasi, R. and Dhanya, C. T., 2016, "Changing characteristics of extreme wet and dry spells of Indian monsoon rainfall",J. Geophys. Res., 121, 5, 2146-2160, doi: 10.1002/2015JD024310.

6

Climate-Smart Small Ruminant Production

A Sahoo

Introduction

Climate-smart small ruminant production is an approach for transforming and reorienting the small ruminant production under the new realities of climate change. The main objective of climate-smart small ruminant production is to make the production sustainable for national food security, enhancement of resilience, and reduction of greenhouse gasses emission. Climate-smart agriculture is that, which increases the sustainable productivity, enhances resilience, reduce greenhouse gasses (GHG) and boost the achievements of national food security and development goals (FAO, 2013). In India, most of the rural communities earn their livelihood on smallholder livestock production system and they are very much vulnerable to climate change. Therefore, the need of the hour is to address the impact of climate change both in terms of adaptation as well as mitigation perspective (Vemeulen *et al.*, 2013). The increase in population, the higher income, urbanization, and change in dietary preference leading to increased demand for animal products that pressurizing for higher productivity of livestock (Delgado *et al.*, 1999; Thronton *et al.*, 2007).

The mutton production in our country was 399 million Kg in 2012 which will be around 537 million Kg by 2020, and 840 million Kg in 2030 and 1317 million Kg in 2050. However, the requirement for mutton would be 813 million Kg by 2020, 986 million Kg by 2030 and 1408 million Kg in 2050. Similarly, the wool production in the country is 44.7 million Kg in present time (2011-2012), which have to be increased 150, 180 and 200 million Kg in 2020, 2030 and 2050, respectively. But the harsh climatic condition, shrinkage grazing resources, increase in cultivation and industrialization and decline interest of new generation for small ruminant rearing; making a question mark to meet the future demand in the current production trend. The scarcity of resources, impact of climate change and increase demand for mutton has made the traditional coping mechanism less effective (Sidahmed, 2008). Therefore, we need climate-smart sheep and goat production option that can achieve the triple win scenario of increasing productivity, adapting and building resilience to climate change through a reduction of greenhouse gas (GHG) emission (FAO, 2013; Shikuku *et al.*, 2016).

The impact of climate change in small ruminant production is worldwide because of their dependence on grazing in an extensive system of production. The impact is more severe in countries like India, where climatic variability throughout the country is enormous. In an intensive system of production, the impact is less severe and indirect like the impact of lower crop yield, feed scarcity, and higher energy price will increase the production cost. In general, other impacts are the emergence of livestock diseases, higher temperature and changing rainfall pattern, alter the abundance, and distribution and transmission of pathogens (Baylis and Litheko, 2006).

Livestock contributes about 18% to the global anthropogenic GHG emissions. accounting for 37% of anthropogenic methane and 65% of anthropogenic nitrous oxide (FAO 2006). The trend in enteric methane emission (EME) countries in table 1. The increase of GHG emissions by Indian livestock was more than the world (74% vs 54.3%) but less (74% vs 82%) than the developing countries variability 1961 to 2010. With this trend, world GHG emissions could reach $3{,}520\times10^{9}$ kg CO_2-eq by 2050 due to animal population growth was driven by increased demands for meat and dairy products in the world (Patra 2014).

Table 1: Trends in enteric methane emission from livestock (*Source*: Patra, 2014)

	Global livestock	Indian livestock
Increase in EME from 1961-2010	54.3%	74%
Projected increase in EME by 2050	$120\text{x}10^{9}$ kg	$18.8\text{x}10^{9}$ kg
Annual Growth rate in EME	Goat (2.0%) Buffalo (1.55%) Swine (1.53%)	Goat (1.91%) Buffalo (1.55%) Swine (1.28%) Sheep (1.25%) Cattle (0.70%)

Pillars of climate-smart small ruminant production

a. Productivity: Sustainably increasing productivity without any negative impact on the environment which will provide better food and nutritional security. Productivity enhancement refers to sustainable intensification.

b. Adaptation: For maintaining productivity, adaptation to the changing the climate is essential. It strengthens the resilience by building capacity to adapt to short or long term stresses. Different adaptation options are available which include technological options, behavioral modifications, and managerial choice and policy alternatives.
c. Mitigation: Mitigation indicates to reduce the GHG emission for each kilo of mutton, a fiber that we produce. The production system should manage soil and tree in such a way that maximizes their potential to act as carbon sinks and absorb CO_2 from the atmosphere.

Strategies in climate-smart small ruminant production

The option for climate-smart production system may work with either mitigation and adaptation synergistically or mitigation only or adaptation only.

Grazing management: Rotational grazing can be a useful strategy for grazing management in climate-smart sheep and goat production. Rotational grazing gives the option to adjust the frequency and timing to livestock grazing needs. Rotational grazing allows the maintenance of forage at an earlier growth stage, which provides better quality digestibility as well as better productivity and reduces city emission per unit of live weight gain (Eagle *et al.*, 2017). The migration is a traditional strategy of nomadic farmers in the arid and semi-arid region. This system increases the resilience of the production system under changing the climate. In the intensive system, cultivation of improved varieties of pasture by replacing some native grasses with high yielding and more digestible forages, perennial fodders, pastures, and legumes may be an effective pasture management option in climate-smart sheep and goat production (Bentley *et al.*, 2008).

Animal breeding: selecting higher productive animals to enhance productivity and thereby reduction of CH_4 emission intensity may be a breeding strategy (Waghorn and Hegarty, 2011). Cross-breeding can provide adaptation, food security, and mitigation benefits. Preferably cross-breeding with locally available adapted breeds, which are tolerant to heat stress, nutritional stress, parasites, and diseases are going to be an important component of the climate-smart sheep production system.

Feed resource and fodder bank: The sheep and goats generally graze on pastures, wastelands, fallow lands. However, during the lean period, the feed and fodder scarcity occurs. To meet the demand during that period top feed resources can be fed to the small ruminants. The trees and shrubs of different agroclimatic zones having significance in small ruminant feeding. The country tree leaves are harvested and sun-dried at the appropriate stage and stored, for use as supplements in addition to grazing during the lean period of the region. As looping is prohibited, the dry fallen leaves from the surface, which is almost estimated that 300 to 350 million MT and the grass is available from the forest with good nutrient content can be used to fed during scarcity period. Furthermore and most importantly, the unconventional feed eg. Monsoon Herbage, grasses, and fodder may be a good source of feed during the scarcity period. The unconventional feed can be stored as fodder bank by making good quality silage, feed pellet, feed blocks. The silage may be feed to the animal during the scarcity of water and food. It can be effective in improving feed quality in extream climate change area.

Reproduction management: Although breeding takes place throughout the year, most breeding is linked with highly seasonal availability of grazing resources. The rams and bucks stay with the flock throughout the year, but sheep flock owners, especially in Rajasthan and Gujarat tie the prepuce with a cotton tape so as to avoid matings during undesirable seasons. Most sheep breeding takes place in July–August i.e. immediately after the onset of the monsoon and some of it in March–April, when stubble grazing and Acacia and *Prosopis* pods are available to the animals. Age at first mating in sheep is nearly 1 year. Sheep generally lamb only once in a year (Acharya, 1985). Research at ARC, Bikaner, revealed that 80 to 100% of animals exhibit oestrus throughout the year. However, considering lambing percentages and lamb survival and growth, breeding in March-April and August-September is preferable (Acharya, 1982). Furthermore, to increase the sustainability of production, its time to get 3 crops in two years. For this, artificial reproductive techniques like estrus synchronization and artificial insemination may be done in the flocks avoiding the lean period for lambing can improve the production of the flock.

Health Management: Global climate change shifts the efficiency and transmission pattern of hosts. With the change of atmospheric humidity and temperature the proliferation of the pathogen and vector. Altogether, the climate alters the occurrence, pathogenicity, spread, transmission, timing, and intensity of outbreaks. The exotic diseases, the main effect of climate change may be an increase in the geographic range and an increase in competence of the non-vertebrate vector. Consideration on the rights of indigenous, migratory and pastoral people in the formulation of strategies, need to develop a comprehensive plan (e.g. health, disaster reduction) to deal with the migration of disease due to climate change. A positive animal welfare contingency plans to control zoonoses caused by climate change has to be developed. Use of vaccinations as a preventive measure where appropriate in regions where the disease is endemic. The transportation of live animals has been limited. Finally, by combining improved empirical data and refined models with a broad view of the small ruminant production system, robust projections of disease risk can be developed.

Vaccines: Vaccination against methanogens in the rumen can be a potentially effective mitigation option for ruminants, which also reared in the low-input extensive system (Wright and Kleive, 2011).

Weather forecasting and Insurance: The proper information on the weather forecast for the rural communities will help to manage the risk as well as taking adaptive measures to prevent climatic variability. In such situations, livestock insurance schemes which are weather-indexed (i.e. policyholders are paid in response to trigger events such as abnormal rainfall or high local animal mortality rates) may be effective where preventive measure fails (Skees and Enkh-Amagala, 2012).

Agro-forestry: Agroforestry is important to climate change mitigation by contributing to carbon sequestration, improved feed and consequently reduces enteric methane. Tree shades provide protection against heat stress during extreme summer. Trees also provide quantity forage that reduces overgrazing and land degradation (Thornton and Herrero, 2010).

Water management: In near future water management is going to be a critical factor for any type of production system. The practice of increasing the output per unit of water use will be considered as an adaptive and climate-resilient production system. A number of adaptive techniques and approaches are available to increase the water use efficiency for crop production that indirectly influences the water management of livestock production.

Feed conversion: Higher feed conversion ratio reduces the amount of feed requires per unit of animal product. Feed efficiency can be increased by selecting breeds of faster growth rate, handier, feed efficiency can also be improved by improving herd health through better veterinary services, preventive health programs, and improved water quality. Selecting feed with a low carbon footprint is another way to reduce emissions.

Building resilience along the supply chain: climate change increases the price volatility of inputs, the feed and energy that increases the financial risk for stakeholders involved in the livestock supply chain. The changing disease pattern due to changing climate further added to the risk. Probably greater coordination among the different stakeholders involved in the supply chain, insurance schemes, buffers and stocks may contribute to greater resilience of supply chain that relays on the landless livestock system.

Conclusion

Sheep husbandry is a sustainable livelihood option for landless marginal farmers, especially arid and semi-arid region since ancient time. Although sheep husbandry itself is climate-smart agriculture, still there is a scope to make it better climate-smart sheep husbandry under the pressure of changing the climate and increasing demand. Alternative strategies have to be implemented in sheep husbandry practices especially during the scarcity period of feed and water. To make the production system more sustainable nutritional intervention, water management strategies have to be initiated. All the efforts should be directed towards sustainable climate-smart agriculture to provide a better livelihood to the poor farmers and protecting the climate.

References

Baylis M and Githeko AK 2006. The effects of climate change on infectious diseases of animals. Report for the Foresight Project on Detection of Infectious Diseases, Department of Trade and Industry, UK Government.

Bentley D and Hegarty R 2008. Managing livestock enterprises in Australia's extensive rangelands for greenhouse gas and environmental outcomes: a pastoral company perspective. Australian Journal of Experimental Agriculture 48, 60-64.

Delgado C, Rosegrant M, Steinfeld H, Ehui S, Courbois C 1999. Livestock to 2020: The next food revolution. Food, Agriculture, and the Environment Discussion Paper 28, International Food Policy Research Institute, Washington, DC.

Eagle AJ, Olander LP, Henry LR, Haugen-Kozyra K, Millar N, Robertson GP 2012. Greenhouse gas mitigation potential of agricultural land management in the United States: a synthesis of literature. Report NI R10-04, Third Edition. Durham, USA, Nicholas Institute for Environmental Policy Solutions, Duke University.

FAO 2006. Livestock's long shadow. Environmental Issues and Options Food and Agriculture Organization of the United Nations; Rome, Italy

FAO 2013. Climate-Smart Agriculture. Food and Agricultural Organization of the United Nations. Sourcebook, Rome, Italy.

Hoffman MT and Vogel C2008. Climate change impacts on African rangelands. Rangelands 30, 12–17.

Patra AK 2014. Trends and Projected Estimates of GHG Emissions from Indian Livestock in Comparisons with GHG Emissions from World and Developing Countries Asian-Australasian Journal of Animal Sciences, 27: 592-599.

Shikuku KM, Mwongera C, Winowiecki L, Twyman J, Atibo C, Läderach P 2015. Understanding Farmers' Indicators in Climate-Smart Agriculture Prioritization in Nwoya District, Northern Uganda. International Center for Tropical Agriculture (CIAT), Cali, Colombia 46 p.

Sidahmed A 2008. Livestock and climate change: coping and risk management strategies for a sustainable future. Livestock and Global Climate Change conference proceeding, Tunisia.

Skees JR and Enkh-AmgalaA 2002. Examining the feasibility of livestock insurance in Mongolia. Policy Research Working Paper 2886, Washington DC, World Bank.

Thornton P, Herrero M, Freeman A, Mwai O, Rege E, Jones P, McDermott J 2007. Vulnerability, Climate change and Livestock – Research Opportunities and Challenges for Poverty.

Thornton PK and Herrero M 2010. Potential for reduced methane and carbon dioxide emissions from livestock and pasture management in the tropics. New York, USA, PNAS.

Vermeulen SJ, Challinor AJ, Thornton PK, Campbell BH, Eriyagama N, Vervoort JM, Kinyangi J, Jarvis A, Laderach P, Ramirez-Villegas J, Nicklin KJ, Hawkins E, Smith DB 2013. Addressing uncertainty in adaptation planning for agriculture. Proceedings of the National Academy of Sciences 110, 8357–8362.

Waghorn GC and Hegarty RS 2011. Lowering ruminant methane emissions through improved feed conversion efficiency. Animal Feed Science Technology 166–167, 291–301.

Wright ADG and Klieve AV 2011. Does the complexity of the rumen microbial ecology preclude methane mitigation? Animal Feed Science Technology 166–167, 248–253.

7

Physiological Response of Cattle to Climatic Stress

S R Mishra

Introduction

In the past few decades, persistent change in climatic variables coupled with global warming lead to the genesis of climatic stress. Currently climatic stress is considered as the most serious ultimatum to livestock's growth, development, production and reproduction in tropics and subtropics including India. Livestock's in tropical countries are most commonly affected by the menace of climatic stress. It has been shown that, climatic stress upsets animal's body homeostasis resulting in grievous decline in livestock's production and productivity across the world. In general, livestock adapt to the climatic insults via behavioural, physiological, biochemical, metabolic, endocrine and molecular responses. Despite of these thermal adaptation mechanisms, livestock's witnesses the perils of climatic stress thereby incur major loss in their production and productivity. Therefore, in this changing climatic scenario, it is imperative to rigorously understand and explore the precise mechanism of thermal adaptation to generate climate resilient species which could upsurge the socio-economic status of the farmers as well as the country.

Physiological responses to climatic stress

Physiological responses are the immediate response shown by livestock against climatic stress and categorized as rectal temperature, respiration rate, heart rate, sweating rate, skin temperature along with digestion and absorption of nutrients. Physiological responses also include endocrine repertoire as alternation in hormonal levels play critical role in livestock's adaptation to climatic stress. The animals respond to hyperthermia via these physiological and endocrine responses to maintain their body homeostasis by balancing the heat gain and heat loss from their body to the environment. Hence, this chapter briefly describes about the physiological and endocrine responses in which helps in acquisition of thermo-tolerance to livestock's.

Rectal temperature

Rectal temperature is regarded as sensitive indicator of physiological status during climatic stress in domestic species. Species, breed, age, sex, nutrition, feeding time, body condition, feeding management, previous heat exposure, shelter management and cooling strategies regulates rectal temperature during climatic stress. Even 1°C increase in ambient temperature beyond the upper critical temperature result in severe decline in livestock's production and productivity. Various reports showed that, rectal temperature increases with increase in ambient temperature in domestic species. In Karan Fries cattle, the rectal temperature was found to be highest in peak afternoon and lowest during midnight in different seasons (Vaidya *et al*., 2011). It was reported that, the rectal temperature was increased in cattle upon heat exposure. Chaiyabutr *et al*. (2008) found higher rectal temperature during peak day time in crossbred Holstein Friesian cattle. It was also reported that, the rectal temperature increased from 38.6°C to 40.4°C on heat exposure in lactating dairy cows. In another study in crossbred cattle, the rectal temperature was up-regulated upon long term heat exposure for 25 days in psychometric chamber. Normally, *Bos indicus* cattle are more resistant to chronic heat exposure as compared to *Bos taurus* due to their lower rectal temperature during peak summer. The hike in rectal temperature during climatic stress could disperse the excessive heat waves to maintain animal's body homeostasis.

Respiration rate

Respiration rate is considered as most consistent biomarker amongst the different physiological responses shown by livestock's against climatic stress. An increase in respiration rate was noticed in cattle during heat stress. Upon thermal stress, the respiration rate was significantly increased in the lactating dairy cows. Likewise, higher respiration rate was observed in Angus cattle during acute heat stress. The respiration rate was found to be increased with increase in temperature humidity index (THI) in Sahiwal and Karan Fries cattle (Sailo *et al*., 2017). Additionally, the respiration rate was found to be highest at peak afternoon and lowest at midnight in karan fries cattle among different seasons (Vaidya *et al*., 2011). It was reported that, the respiration rate was up-regulated after day 1 of heat exposure and subsequently decreased in chronic heat exposure in young calves. Similarly, thermal stress resulted in higher respiration rate in lactating Holstein Friesian cows compared to cows in thermo-neutral zone. Above all, higher respiration rate was noted during late afternoon than morning hours in crossbred calves during summer stress. Nonaka *et al*. (2008) demonstrated that, respiration rate increased up to 3 fold during high ambient temperature. Parihar *et al*. (1992) documented significant change in respiration rate between the cattle present in open environment than those kept under shed. Likewise, the respiratory frequency was found to be higher in free ranged cattle than those kept under shed. Similarly, Pereira *et al.* (2008) revealed that, the respiratory frequency doubles in different breeds of cattle during late afternoon. The increase in respiration rate may induce evaporative heat loss thereby causes cooling to the animals during climatic stress. Moreover, the increase in respiration rate may avert hike in rectal temperature to maintain the body homeostasis.

Heart rate

Variation in heart rate and pulse rate are appraised as indicator in livestock's against climatic stress. During climatic stress, rise in the peripheral catecholamines level lead to increase in heart rate in cattle. The heart rate was known to be increased during acute heat stress in comparison with chronic heat stress. It has been investigated that, *Bos taurus* possess higher heart rate than *Bos indicus* at 32°C ambient temperature. It has also been seen that, increase in environmental temperature increases pulse rate in cattle. In Karan Fries cattle, Vaidya *et al.* (2011) revealed highest pulse rate during peak afternoon and lowest during late night in different seasons. In contrast, few studies reported that, the heart rate was declined with increase in heat exposure (Singh and Newton, 1978). However, the heart rate did not differ between summer and winter season in Holstein Friesian cattle. Similarly, the heart rate did not alter significantly as like rectal temperature and respiration rate against climatic stress in the primiparous Holstein Friesian and Jersey cows (Muller and Botha, 1993).

Sweating rate

Sweating is considered as a vital process in dairy cattle during climatic stress as it eliminates excessive heat from the animal's body. Various environmental variables modulate sweating rate such as ambient temperature, relative humidity, wind velocity and solar radiation. Apart from these, several factors control the cooling to the animal's body and the intensity of evaporative heat loss such as hair, density and thickness of hair coat and skin colour. Thus, the evaporative heat loss seems to be the major process of cooling in cattle exposed to high environmental temperature. Hillman *et al.* (2001) elucidated that, dark skinned dairy cows have greater sweating rate as compared to light skinned dairy cows. This report was further confirmed by the same group of researchers in heifers where dark skinned heifers gained more heat than light skinned heifers.

Skin temperature

The skin and cutaneous circulation confer the heat exchange between the body and the surrounding environment. Generally, the skin temperature was increases with increase in ambient temperature. It has been shown that, the skin temperature increased by 0.22°C with every 1°C rise in ambient temperature in domestic animals (Paulo and Lopes, 2014). The increase in skin temperature might render vasodilation of cutaneous capillaries thereby causes rushing of the blood flow to the peripheral circulation to enhance excessive heat loss.

Digestion and absorption of nutrients

Climatic stress affects all the factors responsible for digestion i.e. feed quality, rate of feed intake, composition of feed, feed palatability, rate of feed passage and capacity of rumen and post-gastric compartments. It is obvious that, feed intake reduces during climatic stress thus rate of digestion increases due to slow passage of feed materials into digestive compartments. On the other hand, blood flow to peripheral organs increases while blood flow to visceral organs decrease during heat stress. This may decrease the absorption nutrient as a result of reduction in blood supply to portal vein.

Endocrine responses

It is well known that, the neuro-endocrine system plays a key role in thermal adaptation of livestock against climatic stress. Basically, acute heat stress activates hypothalamo-pituitary adrenal (HPA) axis and sympathetic adrenal medullary (SAM) axis to alter the secretion of various hormones to rescue the livestock from thermal insults. Several hormones are released into the systemic circulation to accomplish the task of thermal adaption in cattle. Those hormones are namely catecholamines, cortsiol, gonadotropin releasing hormone (GnRH), follicle stimulating hormone (FSH), luteinizing hormone (LH), thyroid hormone, insulin, growth hormone (GH), insulin like growth factor-I (IGF-I), aldosterone, prolactin and anti diuretic hormone (ADH) etc.

It has been well established that, acute exposure to high ambient temperature activates the HPA and SAM axis to release catecholamines and cortisol in all domestic mammals. Interestingly, cortisol is considered as the primary stress hormone in cattle. The cortisol level was found to be elevated in response to acute heat stress and declined during chronic heat stress in cattle. Solar radiations during summer stress resulted in hike of cortisol level from 11 to 29 ng/ml in Holstein Friesian calves (Yousef *et al.*, 1997). In another study conducted in psychometric chamber, the cortisol level was surged from 3.8 to 6.5 ng/ml when animals were exposed to increase in temperature from 24 to

38°C for 9 hours (Habeeb *et al.*, 2001). In Holstein Friesian claves, the cortisol level was reduced by 45% during long term heat exposure (Kamal *et al.*, 1989). This may be due to the negative feedback of glucocorticoid upon HPA axis and reduction in cortisol binding globulin or transcortin. Moreover, cortisol regulates different body metabolism and serve as vasodilator to stimulate heat dissipation thereby assist the animals to counteract the deleterious effects of heat stress. In cattle, the catecholamines were found to be increased in short term as well as long term heat stress. This finding suggests that, higher level of catecholamines may provoke sweat glands to maintain thermal balance.

High environmental temperature averts hypothalamic GnRH neurons to suppress the release of cascade of molecules such as GnRH, FSH and LH thereby affects reproduction in livestock's species. The reduction in the aforementioned hormones dwindle the steroidogenic potential of sertoli cells and granulosa cells. It not only reduces testosterone level but also declines the spermatogenesis. Subsequently, the semen volume, semen quality, number of spermatozoa, number of progressively motile spermatozoa and spermatozoa concentration were found to be reduced during summer stress. As a consequence, the number of abnormal spermatozoa and aged spermatozoa got increased during summer stress compared to winter and spring. Likewise, in females, lower circulating levels of 17β-estradiol depress oocyte competence and ovulation leading to early embryonic mortality. In another study, lower circulating level of LH and 17β-estradiol resulted in summer infertility in high yielding dairy cows. Consequently, the conception rate was reduced around 20-27% in dairy cows during peak summer (Chebel *et al.*, 2004). In addition, higher cortisol level down regulates GnRH as well as LH to preclude reproductive cyclicality of domestic animals. It has been reported that, climatic stress declines pregnant estrogen or estrone sulfate thereby lowers the birth weight of neonatal calves. During climatic stress, the PGF2α secretion was increased from uterine endometrial cells to cause luteolysis culminating in pregnancy failure.

Thyroid hormones do play vital role in reproductive performance of livestock's species. It was reported that, the thyroid gland is highly sensitive to variation in environment temperature. During extreme ambient temperature, the activity of thyroid gland is reduced resulting in hypothyroidism and low metabolic heat generation. The concentration of tri-iodothyronine (T_3) and thyroxine (T_4) were decreased in dairy cattle in response to climatic stress. In cattle, Farooq *et al.* (2010) observed lower level of thyroid hormone during summer season compared to winter. Earlier study in cattle showed a dip in T_3 level from 2.2 to 1.16 ng/ml during heat stress. The level of insulin was declined by 54% in Holstein Friesian heifers, 33% in Holstein Friesian cows and 30% in

Holstein Friesian calves exposed to climatic stress (Marai and Haeeb, 2010). This could be explained by the fact that, the negative energy balance during thermal stress may reduce the secretion of insulin into peripheral circulation. In Jersey cows, the GH level was found to be 18.2 ng/ml in thermo-neutral zone (TNZ) and 13.5 ng/ml when exposed to heat stress. This decline in GH level could suppress calorigenesis as well as body metabolism to maintain the body homeostasis during thermal stress. The IGF-1 was also noted to be decreased in cattle during climatic stress. Moreover, the GH stimulates the thyroid hormone secretion from thyroid gland. Thus, decline in GH might reduce the thyroid hormone secretion thereby dwindle body calorigenesis to combat the lethal effects of climatic stress.

It is interesting to know that, corticotropin releasing hormone (CRH) triggers growth hormone inhibiting hormone (GHIH) or somatostatin to decline the GH and thyroid hormone levels during climatic stress. In cattle, the aldosterone level remains normal during acute heat stress while reduces by 40% after chronic heat exposure. The reduction in plasma aldosterone level could be attributed to loss of potassium ions via excessive sweating during chronic heat stress. In Holstein Friesian heifers, the prolactin level was spiked by 3 fold when ambient temperature increased from 18 to 32°C (Alamer, 2011). The cause for increase in prolactin level has not yet understood. Still, higher prolactin level may expedite the insensible heat loss and activate sweat gland during climatic stress. However, another group of researchers explained that, higher prolactin level may control the water and electrolytes balance during climatic stress. In contrast, prolactin level was also decreased upon heat exposure in ruminants (Bocquier *et al.*, 1998). Most of the domestic mammals undergo severe dehydration on exposure to climatic stress. Thus, during extreme summer, the ADH secretion increases from posterior pituitary and takes care of the plasma osmolarity by re-absorption of water from renal tubule.

Conclusion

Till date, climatic stress seems to be very pernicious in livestock's production and productivity throughout the world. Cattle do respond to the deleterious effects of climatic stress via different physiological and hormonal mechanisms. Thus the rectal temperature, respiration rate, heart rate, sweating rate, skin temperature and the circulating concentrations of various hormones could be taken as ideal biological markers to quantify the intensity of climatic stress in livestock's. Future research works are warranted on cellular and molecular studies which could to generate climate resilient domestic species for sustainable livestock's production across the world.

References

Alamer M. 2011. The role of prolactin in thermoregulation and water balance during heat stress in domestic ruminants. Asian Journal of Animal and Veterinary Advances, 6: 1153-1169.

Bocquier F, Bonnet M, Faulconnier Y, Guerre-Millo M, Martin P and Chilliard Y. 1998. Effects of photoperiod and feeding level on adipose tissue metabolic activity and leptin synthesis in the ovariectomized ewe. Reproduction Nutrition Development, 38: 489-498.

Chaiyabutr N, Chanpongsang S and Suadsong S. 2008. Effects of evaporative cooling on the regulation of body water and milk production in crossbred Holstein cattle in a tropical environment. International journal of biometeorology, 52: 575-585.

Chebel RC, Santos JEP, Reynolds JP, Cerri RLA, Juchem SO and Overton M 2004. Factors affecting conception rate after artificial insemination and pregnancy loss in lactating dairy cows. Animal Reproduction Science, 84: 239-255.

Farooq U, Samad HA, Shehzad F and Qayyum A. 2010. Physiological Responses of Cattle to Heat Stress. World Applied Sciences Journal, 8: 38-43.

Habeeb AAM, Aboulnaga AJ and Kamal TH. 2001. Heat-induced changes in body water concentration, Ts, cortisol, glucose and cholesterol levels and their relationships with thermoneutral body weight gain in Friesian calves. Proceedings of 2nd International Conference on Animal Production and Health in Semi-arid Areas, pp 97-108. El-Arish, North Sinai, Egypt.

Hillman PE, Lee CN, Carpenter JR, Baek KS and Parkhurst A. 2001. Impact of hair color on thermoregulation of dairy cows to direct sunlight. In: ASAE Annual Meeting 1998. American Society of Agricultural and Biological Engineers.

Kamal TH, Habeeb AA, Abdel-Samee AM and Marai IF. 1989. Milk production of heat stressed friesian cows and its improvement in the subtropics. European association of Animal Production, 38: 156-158.

Marai AAM and Haeeb B. 2010. Buffalo's biological functions as affected by heat stress-A review. Livestock Science, 127: 89-109.

Muller CJC and Botha JA. 1993. Effect of summer climatic conditions on different heat tolerance indicators in primiparous Friesian and Jersey cows. South African Journal of Animal Science, 23: 98-103.

Nonaka I, Takusari N, Tajima K, Suzuki Higuchi T and Kurihara KM. 2008. Effect of high environmental temperatures on physiological and nutritional status of prepubertal Holstein heifers. Livestock Science, 113: 14-23.

Parihar AS, Jain PK, Singh HS, Singh YP and Singh NP. 1992. Seasonal variations in physiological responses of crossbred calves under different housing system. Indian Journal of Animal Science, 62: 686-688.

Paulo JLA and Lopes FA. 2014. Daily activity patterns of Saanen goats in the semi-arid northeast of Brazil. Revista Brasileira de Zootecnia, 43: 464-470.

Pereira AMF, Baccari Jr F, Titto EAL and Almeida JAA. 2008. Effect of thermal stress on physiological parameters, feed intake and plasma thyroid hormones concentration in Alentejana, Mertolenga, Frisian and Limousine cattle breeds. International Journal of Biometeorology, 52: 199-208.

Sailo L, Gupta ID, Das R and Chaudhari MV. 2017. Physiological Response to Thermal Stress in Sahiwal and Karan Fries Cows. International Journal of Livestock Research, 7: 275-83.

Singh SP and Newton WM. 1978. Acclimatization of young calves to high temperatures: physiological responses. American Journal of Veterinary Research, 39: 795-797.

Vaidya MM, Kumar P and Singh SV. 2011. Circadian changes in heat storage and heat loss through sweating and panting in Karan Fries cattle during different seasons, Biological Rhythm Research, 43: 137-146.

Yousef JLM, Habeeb AA and EL-Kousey H. 1997. Body weight gain and some physiological changes in Friesian calves protected with wood or reinforced concrete sheds during hot summer season of Egypt. Egyptian Journal of Animal Production, 34: 89-101.

8

Genomic Approaches to Identify Heat Tolerance in Livestock

B P Mishra, Amit Kumar and B Sajjanar

Introduction

Animals have delicate balance of heat production and heat loss, which is maintained by thermoregulatory mechanisms in response to environmental temperature and humidity combinations. When the heat loss is overrun by heat gain, heat stress occurs. During heat stress, increase in the core body temperature due to failure of homeostatic mechanism reduces productivity of the animals below their original genetic potential in growth, milk production, milk constituents and reproduction (Ravagnolo and Misztal, 2002). Impending climate change scenario is a unanimously accepted reality. Increase in the anthropogenic greenhouse gases (CO_2, $CH_4$and NO_2) cause radiation leading to excess warming of the earth's environmental system. According to IPCC, the predicted temperature rises for the end of 21st century may be in the range of 1.8°C to 4°C (IPCC, 2007). The climate change has complex impact on the livestock production. Temperature or heat stress will be one of the major environmental factor to influence the health and productivity of livestock population. The least tropically adapted livestock may be severely affected by heat stress. The manage mental and ameliorative practices can deal with the heat stress in short term, however the long term focus of developing genetic heat stress tolerance in livestock population will be a significant and impactful strategy.

Understanding the genetic basis of heat stress response with special emphasis on how genes interact with environmental temperatures may lead to identify important genetic markers for stress resilience. Tropically adapted breeds of livestock may possess genetic makeup to relatively resist temperature stress. Selection of animals for these genetic markers or introgression of desirable genetic elements will lead to resilient livestock population. For example, owing to their long time adaptation with tropical climates, zebu breeds (*Bos indicus*) of cattle are better able to regulate body temperature in response to heat stress than European breeds (*Bos taurus*) (Beatty *et al.*, 2006). To enhance the milk production, cross breeding with high producing Holstein Friesian (up to 75% of the exotic blood) has been followed in our country. This has resulted in new strains of dairy cattle such as *Jersind, Jerthar, Karan-swiss, Karan-fries, Sunandini, Frieswal, Phule-Triveni* and *Vrindavani* which are capable of producing more milk than native breeds. However, these cross bred animals are found more susceptible to heat stress than the native breeds. High temperature coupled with the high humidity brings significant decrease in milk production in these cross bred population leading to severe economic losses in dairy industry (Agarwal and Upadhaya, 2013).

There is a need to understand molecular mechanisms and identify genetic markers for heat stress resilience in different livestock population. The genomic approaches with recent high throughput techniques have definite scope in unravelling the genetic basis for heat stress resilience in livestock population to improve their productivity in the impending climate change scenario.

Physiological parameters to measure heat stress tolerance in livestock

Relative heat stress tolerance of livestock determined by the differences in the physiological responses exhibited by the animals during exposure to stress conditions. The physiological responses/parameters to thermal stress include body temperatures, respiration rates, sweating rates etc. Body temperatures including surface (head, dorsal, ventral, rump) and rectal temperatures that can be recorded by clinical and infrared thermometers. Respiration rates(RR) are to be recorded by counting the flank movements from a distance prior to examining the rectal temperature in order to avoid any disturbance to the animals. The relative stress limits in dairy animals in-terms of Heat Tolerance Co-efficient (HTC) calculated by following a standard formula: $HTC = 100 - 10 (ART - 38.3)$, where, HTC is the heat tolerance coefficient, ART is the average rectal temperature, 38.3 is the physiological bovine body temperature, 10 is a correction factor to convert deviations in body temperature to a unit basis and 100 is the perfect efficiency in maintaining temperature at 38.3°C. The measurement of levels of different stress hormones and enzymes such as cortisol, glutathione peroxidase (GSH) and Superoxide dismutase (SOD) provide biochemical indicators of heat stress levels in the affected animals.

Genetic mechanisms of heat stress response in livestock

Cellular responses to heat stress involve highly conserved mechanism of altered gene expression and protein activation. Gene networks within or across the cells and tissues co-ordinate the cellular and whole animal metabolism in dairy cattle (Collier *et al.*, 2008). There are different approaches to study the genetic basis of heat stress tolerance in livestock including candidate gene approach and recent advanced methods that employ omics approaches.

Candidate gene marker approaches

Many candidate genes have been identified as responders to thermal stress. Studies have been carried out to identify allelic variants in some genes that confer heat tolerance in cattle, such as slick hair gene, ATP1B2 gene, HSP70A1A, HSP90AB1, etc. One of those genes is ATP1A1, which encodes for the $\alpha 1$ chain of Na+/K+-ATPase that contains the catalytic unit of the enzyme. The Na+/K+-ATPase is a membrane bound active transport system responsible for maintaining the low internal Na+ and high internal K+ across the plasma membrane that is typical to most animal cells. Na+/K+ATPase alpha chain is expressed in all tissues predominantly in peripheral nerves and in erythrocytes. The importance of Na+/K+ATPase in basal metabolism can be highlighted by the fact that it consumes 19-28% of total ATP production in mammalian cells at rest, to actively transport 3 Na+ of and 2 K+ into the cell. It has been established that Indian breeds of cattle are better able to regulate their body temperature in response to heat stress than the exotic breeds of cattle. Further, increase in exotic blood level in crossbreds leads to depression of production potential during heat stress. However, involvement of few genes cannot increase animal's ability to dissipate heat and the same may require coordination of multiple genes at higher regulatory control system. This coordinated process may involve programmed expression of genes that supports improved thermo-tolerance. The identification of these genes and their control system is possible through modern technological advances. The recent 'Omics' approaches (phenomic, genomic and transcriptomic) may enable to understand the genetic component of heat tolerance in livestock population.

Transcriptomic approaches

The complexity of heat shock response indicates involvement of different mechanisms that are genetically regulated. Identification of these multiple genes associated with thermal stress is possible with application of genome-wide expression techniques such as microarray or RNA sequencing methods. Recently studies are undertaken to analyse gene expression patterns during heat stress in different livestock species such as cattle, pig, horse and poultry. The results enable the identification of those genes that are involved in key regulatory pathways that supports heat stress response. Transcriptomic analysis in heat stressed dairy cattle identified groups of genes that are involved in increased glucose and amino acid oxidation and reduced fatty acid metabolism, endocrine system activation etc. Apart from this, genes associated with the skin properties and immune system are also found associated with heat stress response (Kolli *et al.*, 2014; Li *et al.*, 2015). Apart from genes, non-coding RNA have been found to involve in heat stress response. The differentially expressed microRNAs were identified in the heart, liver, kidney, and lung of rats at three time points: during heat stress (i.e., when core temperature reached 41.8 °C), or following a 24 or 48 h recovery period. Global microRNA and transcriptomics analysis suggested that perturbed miRNA due to heat stress are involved in biological pathways related to organ injury, energy metabolism, the unfolded protein response, and cellular signalling. In a study, RT-qPCR and deep-sequencing methods showed that 8 miRNAs among the 12 selected miRNAs (miR-19a, miR-19b, miR-27b, miR-30a-5p, miR-181a, miR-181b, miR-345-3p, and miR-1246) were highly expressed in the serum of heat-stressed Holstein cows. An investigation was performed for mRNA and microRNA (miRNA) expression profiling between heat tolerant (HT, n = 4) and non-heat tolerant (NHT, n = 4) buffaloes, for identifying the specific modules, significant genes, and miRNAs related to the heat tolerance identified using the weighted gene co-expression network analysis (WGCNA). The results indicated that the buffaloes in HT had a significantly lower rectal temperature (RT) and respiratory rate (RR) and displayed a higher plasma heat shock protein (HSP70 and HSP90) and cortisol (COR) levels than those of NHT buffaloes. Total 158 DEGs associated with heat tolerance in the turquoise module were identified, 35 of which were found within the protein-protein interaction network.

Genome Wide Association Studies (GWAS)

Genome Wide Association Studies (GWAS) would be a powerful tool for study of heat tolerance selection in animals. There is a need for genotyping large of number of animals for identification of genome wide SNP markers and DNA markers (commercially available SNP chips) and collecting phenotypes of heat stress response in livestock. Core body temperature during heat stress is a heritable trait in dairy cattle, with early estimates varying from 0.15 to 0.31 and a more recent estimate indicating a value of 0.17. Reliability of genetic estimates for rectal temperature, such as for other genetically-controlled traits, should be improved by genome wide association studies (GWAS) to identify SNPs associated with regulation of rectal temperature. Quantitative trait loci (QTL) can be identified for lowly-heritability traits and used to improve reliability of genetic estimates despite the gain in reliability being less than for more heritable traits. In a reported study, records on afternoon rectal temperature where the temperature-humidity index was more than 78.2 were obtained from 4,447 cows sired by 220 bulls, resulting in 1,440 useable genotypes from the Illumina BovineSNP50 BeadChip with 39,759 SNP. The largest proportion of SNP variance (0.07 to 0.44%) was explained by markers on *Bos taurus* autosome (BTA) 24 in a region flanked by U1 and NCAD. In addition, the SNP on BTA 16 explained 0.08% and 0.11% of the SNP variance. The contig includes SNORA19, RFWD2 and SCARNA3. Other SNPs associated with RT were located on BTA 16 (close to CEP170 and PLD5), BTA 5 (near SLCO1C1 and PDE3A), BTA 4 (near KBTBD2 and LSM5), and BTA 26 (GOT1, a gene implicated in protection from cellular stress). QTLs have been reported for RT in heat-stressed dairy cattle. These SNPs could prove useful in genetic selection and for identification of genes involved in physiological responses to heat stress. In addition, GWAS can be useful for understanding the underlying biology of a trait by identifying candidate genes in physical proximity to QTL. These studies can detect potential genomic signals strongly associated with the heat stress response of animals. The use of new phenotypes in GWAS studies together with genome wide comparison of DNA sequences from tropical and temperate climate breeds could provide valuable and complementary information from the DNA analyses.

Epigenetic mechanisms

The genomic variations are able to explain portion of different phenotypic traits including stress response and the remaining part is considered to be embedded in the epigenome & its dynamic interactions with the environmental factors. Epigenome is theensemble of chromosome modifications (DNA methylation and histone modification) without changing nucleotide sequence of the DNA. Recently there has been considerable evidence that epigenetics mediates the effects of environmental stressors on phenotypes in plants and animals (Kinoshita and Seki, 2014; Feeney *et al*., 2014). Epigenetic changes are considered as source for inheritance of phenotypic variations in livestock (Triantaphyllopoulos *et al*.,2016). Pigs exposed to heat stress were found to have changes in the genome-wide DNAmethylation profiles compared to the control animals kept in thermo-neutral conditions (Hao *et al*., 2016). Prenatal transportation stress was found to alter genome-wide DNA methylation in Brahman bull calves (Littlejohn *et al*., 2018). Recent techniques such as high throughput bi-sulfite sequencing and chromatin immune precipitation sequencing (ChIP-seq) methods are used for identification of epigenetic profiles. The epigenetic mechanisms such as DNA methylation needs to be considered while identification of heat tolerance in livestock.

References

Agarwal A, Upadhay R (2013) Heat stress and hormones. Heat Stress and Animal Productivity. Springer, India, 27-51

Beatty DT, Barnes A, Taylor E, Pethick D, McCarthy M, Maloney SK (2006) Physiological responses of Bos taurus and Bos indicus cattle to prolonged, continuous heat andhumidity. J AnimSci, 84(4): 972-85

Bhat S, Kumar P, Kashyap N, Deshmukh B, Dige MS, Bhushan B, Chauhan A, Kumar A, Singh G (2016) Effect of heat shock protein 70 polymorphism on thermotolerance in Tharparkar cattle, Veterinary World 9(2): 113-117.

Collier RJ, Collier JL, Rhoads RP, Baumgard LH (2008) Invited review: genes involved in the bovine heat stress response. J Dairy Sci, 91(2): 445-454

Dikmen S, Cole JB, Null DJ, Hansen PJ (2013) Genome-Wide Association Mapping for Identification of Quantitative Trait Loci for Rectal Temperature during Heat Stress in Holstein Cattle. PLoS ONE 8(7): e69202. doi:10.1371/journal.pone.0069202

Feeney A, Nilsson E, Skinner MK (2014) Epigenetics and transgenerational inheritance in domesticated farm animals. J AnimSciBiotechnol, 5(1): 48

Garner, J. B. et al. (2016) Genomic Selection Improves Heat Tolerance in Dairy Cattle. Sci. Rep. 6, 34114; doi: 10.1038/srep34114.

Hao Y, Cui Y, Gu X (2016) Genome-wide DNA methylation profiles changes associated with constant heat stress in pigs as measured by bisulfite sequencing. Sci Rep. 6: 27507

IPCC (Intergovernmental Panel on Climate Change), 2007 Summary for policymakers, Cambridge Press, Cambridge, UK, 23

Kashyap N, Kumar P, Deshmukh B, Bhat S, Kumar A, Chauhan A, Bhushan B, Singh G, Sharma D (2015) Association of ATP1A1 gene polymorphism with thermo tolerance in Tharparkar and Vrindavani cattle, Veterinary World 8(7): 892-897.

Kinoshita T, Seki M (2014) Epigenetic memory for stress response and adaptation in plants. Plant Cell Physiol, 55(11): 1859-63

Kolli V, Upadhyay RC, Singh D (2014). Peripheral blood leukocytes transcriptomic signature highlights the altered metabolic pathways by heat stress in zebu cattle. Res Vet Sci, 96(1): 102-111

Li L, Sun Y, Wu J, Li X, Luo M, Wang G (2015) The global effect of heat on gene expression in cultured bovine mammary epithelial cells. Cell Stress Chaperones, 20(2): 381-389

Littlejohn BP, Price DM, Neuendorff DA, Carroll JA, Vann RC, Riggs PK, Riley DG, Long CR, Welsh TH Jr, Randel RD (2018) Prenatal transportation stress alters genome-wide DNA methylation in suckling Brahman bull calves. J AnimSci, 96(12): 5075-5099.

Liu S, Ye T, Li Z, Li J, Jamil AM, Zhou Y, Hua G, Liang A, Deng T and Yang L (2019) Identifying Hub Genes for Heat Tolerance in Water Buffalo (Bubalus bubalis) Using Transcriptome Data. Front. Genet. 10:209. doi: 10.3389/fgene.2019.00209

Misagh Moridi, Seyed Hossein Hosseini Moghaddam, SeyedZiaeddinMirhoseini& Massimo Bionaz (2019) Transcriptome analysis showed differences of two purebred cattle and their crossbreds, Italian Journal of Animal Science, 18:1, 70-79, DOI: 10.1080/ 1828051 X.2018.1482800

Permenter MG, McDyre BC, Ippolito DL, Stallings JD. (2019) Alterations in tissue microRNA after heat stress in the conscious rat: potential biomarkers of organ-specific injury.BMC Genomics. 2019 Feb 15;20(1):141. doi: 10.1186/s12864-019-5515-6.

Ravagnolo O1, Misztal I (2002) Effect of heat stress on nonreturn rate in Holstein cows: genetic analyses. J Dairy Sci. 85(11):3092-100.

Srikanth K, Lee E, Kwan A, Lim Y, Lee J, Jang G, Chung H.(2017) Transcriptome analysis and identification of significantly differentially expressed genes in Holstein calves subjected to severe thermal stress. Int J Biometeorol. 2017 Nov;61(11):1993-2008. doi: 10.1007/s00484-017-1392-3.

Triantaphyllopoulos KA, Ikonomopoulos I, Bannister AJ. (2016) Epigenetics and inheritance of phenotype variation in livestock. Epigenetics Chromatin, 9:31

Zheng Y, Chen KL, Zheng XM, Li HX, Wang GL.(2014) Identification and bioinformatics analysis of microRNAs associated with stress and immune response in serum of heat-stressed and normal Holstein cows. Cell Stress Chaperones. 2014 Nov;19(6):973-81. doi: 10.1007/s12192-014-0521-8

9

Climate Change Impacts and Innovative Adoption Options for Smart Animal-Agriculture

D B V Ramana

Introduction

Climate change has been, and continues to be one of the important causes of low productivity from animal agriculture in tropical countries like India through crop failures, fodder scarcity and increased incidence of endemic animal diseases. It impacts animal agriculture and cause immediate danger for human societies as livelihoods and food production are being adversely affected. Farm ruminant animals contribute to food supply by converting low-value materials, inedible or unpalatable for human, into milk, meat, and eggs and directly contribute to nutritional security. Besides contributing over one-fourth to the agricultural GDP, it provides employment to 18 million people in principal or subsidiary status in India. Nearly two-thirds of farm households are associated with farm animal rearing and 80% of them are small landholders (≤2 ha). As a result of various socio-economic and market driven factors, during the last one decade, large scale transformation took place in both extensive and intensive animal agriculture with regards to type and breed of the animal. A lot of high yielding, low disease resistant and more vulnerable crossbreed animal population have been increased in present animal production systems inorder to meet the market demands. Climate changes could impact severely the economic viability and production of these intensive animal production systems through increased incidence of drought/floods/cyclones/hailstorms etc. Drought and high ambient temperatures in particular, affects production of milk, meat and egg, reproduction, health of animals and condition of pastures. Changes in pasture and crop biomass availability and quality affect animal production through changes in daily or seasonal feed supplies. Heavy rains and associated floods would washout the fodder resources and animals. Recent incidences of severe hail storms are causing brutal injuries to the grazing animals. To mitigate the adverse affects of extreme weather events and cope with changing climate, much precised resilient technologies suitable to local conditions and resources are needed. Hence, one should be critical in recommending smart animal-agriculture technologies in view of much diversified and heterogeneous group of farmers and the resources accessible to them. This will help in sustainable productivity from animal-agriculture and profitability to the farmers even in the era of climate change.

Contribution of Animal Agriculture to Climate Change

The animal agriculture which is vulnerable to climate change is itself a large contributor to global warming. It globally contributes to 7,516 mmt per year of CO_2 equivalent (CO_2 e) GHG emissions through several routes. The most significant are carbon dioxide from land use and its changes (32 per cent), nitrous oxide from manure and slurry (31 per cent) and methane from animal digestion (25 per cent). Estimates of enteric emissions from India vary widely from 6.17 Tg/year to 10.3 Tg/year of which 60% contributed by cattle, 30% from buffaloes and 3% from sheep and 6% from goat and 1% from other species.

Methane traps infrared radiation (IR), or heat, enhancing the greenhouse effect. Its chemically active properties have indirect impacts on global warming as the gas enters into chemical reactions in the atmosphere that not only affect the period of time methane stays in the atmosphere (i.e. its lifetime), but that also play a role in determining the atmospheric concentrations of tropospheric ozone and stratospheric water vapour, both of which are also greenhouse gases. These indirect and direct effects make methane a large contributor, second only to carbon dioxide, to potential future warming of the earth. The global warming potential (GWP) of methane is 21 times more than carbon dioxide. Additionally, methane's chemical lifetime is relatively short, about 12 years compared to 120 years for carbon dioxide.

Nitrous oxide (N_2O) is another potent greenhouse gas, the primary anthropogenic emissions of which are thought to come from agricultural fertilizers, and to a lesser degree, fossil fuel combustion and biomass burning. The GWP of nitrous oxide is 321.

There are two sources of GHG emissions from animal agriculture production systems:

1) From the digestive process (enteric fermentation)
2) From animal wastes (manure)

Climate Change Impacts on Animal Agriculture

The performance, health and well-being of the animals are severely affected by climate both directly and indirectly. The direct effects involve heat exchanges between the animal and its environment that are linked to air temperature, humidity, wind-speed and thermal radiation. These linkages have bearing on the physiology of the animal and influence animal performance (e.g., growth, milk and wool production, reproduction) and health. Hot and humid environmental conditions stress the lactating animal and reduce intake of the nutrients necessary to support milk yield and body maintenance. The primary factors that cause heat stress in milch animals are high environmental temperatures and high relative humidity. In addition, radiant energy from the sun contributes to stress if the animals are not properly shaded. The tremendous amount of body heat generated in high yielding animal is helpful in cold climates but is a severe liability during hot weather. Dry matter intake decreases especially young and productive animals exposed to heat stress. In addition, there can also be a decrease in the efficiency of nutrient utilization and increased loss of sodium and potassium electrolytes. Short-term extreme events (e.g., summer heat waves, winter storms) can result in the death of vulnerable animals, which can have substantial financial impacts on stakeholders. Although the level of vulnerability of the farm animals to environmental stresses varies with the genetic potential, life stage and nutritional status of the animals, the studies unambiguously indicate that the performance of farm animals is directly sensitive to climate factors. Summer weather reduces production of high-producing dairy cows and also the conception rates of dairy cows as much as 36%. With predicted global warming, an additional decline in milk production beyond expected summer reductions may occur particularly in the hot/hot-humid regions. In buffaloes, more silent heats and conception failures were reported. Sudden changes in temperature, either a rise in T max (>4°C above normal) during summer i.e. heat wave or a fall in T min (<3°C than normal) during winter i.e. cold wave cause a decline in milk yield of crossbred cattle and buffaloes. The estimated annual loss at present due to heat stress at the all-India level is estimated at 1.8 million tonnes, that is, nearly 2 percent of the total milk production in the country. Global warming is likely to lead to a loss of 1.6 million tonnes in milk production by 2020 and 15 million tons by 2050 from current level in business as usual scenario (Upadhyay *et al.*, 2007). The decline in yield varies from 10-30% in first lactation and 5-20% in second and third lactation (A.K Srivastava, 2010). Northern India is likely to experience more negative impact of climate change on milk production of both cattle and buffaloes due to rise in temperature during 2040-2069 and 2070-2099. The decline in milk production will be higher in crossbreds (0.63%) followed by buffalo (0.5%) and indigenous cattle (0.4%). A rise of 2-6 °C due to global warming (time slices 2040-2069 and 2070-2099) projected to negatively impact growth, puberty and maturity of crossbreds and buffaloes (Naresh *et al.*, 2012). Time to attain puberty of crossbreds and buffaloes will increase by one to two weeks due to their higher sensitivity to temperature than indigenous cattle. Heat stress induced by climate change has also been reported to decrease reproductive performance in dairy cows. Decrease in weight gain (Ramana *et al.*, 2013) and alterations in reproductive behaviour were also observed in small ruminants. A rise temperature due to global warming projected to negatively impact puberty and maturity in small ruminants. The main effects include decrease in the length and intensity of the oestrus period, decreased fertility rate, decreased growth, size and development of ovarian follicles, increased risk of early embryonic deaths and decreased fetal growth and lamb/ kid birth weight.

Besides the direct effects of climate change on animal production, there are profound indirect effects as well, which include climatic influences on Quantity and quality of feed and fodder resources such as pastures, forages, grain and crop by-residues and the severity and distribution of animal diseases and parasites. A plunge in green fodder availability due to low rainfall is the first major effect of climate change on animal production systems as the fodder crops like sorghum, maize etc fails to grow. Low rainfall causes poor pasture growth from grazing lands and also consequences decline in dry fodder supplies in the form of crop residues. It slows down the plant growth and metabolism as the available soil moisture declines progressively. There is a reduction in cell division and protein synthesis and more lignifications and results in reduction in the total amount of plant biomass and its quality interms of available protein as animal fodder in a given growing season. Rising temperatures increase lignifications of plant tissues and thus reduce the digestibility and the rates of degradation of plant species. Climate change will have a substantial effect on global water availability in future. Not only will this affect drinking water sources for animal agriculture, but it will also have a bearing on animal feed production systems and pasture yield. Further, mechanization in agriculture in recent years leading to 10-15% more reduction in coarse cereal crops availability for animal feeding.

Due to the prolonged inundation during heavy rains and or floods, all the crops including fodder is destroyed or seriously damaged. Wet conditions after floods or heavy rains enhance chances of infection by internal parasites like round worms, tape worms, liver flukes as well as many epidemic diseases like FMD, HS, BQ etc. There may be non-specific water borne infections causing diarrhea and other enteric diseases. Animals that have been standing in water for prolonged periods of time may get pneumonia or foot rot problem.

Lowland sites in relatively low rainfall areas expected reduction of herbage yield a lot in dry seasons. Climate driven models of the temporal and spatial distribution of pests, diseases and weeds have been developed for some key species e.g. the temperate (*Haemaphysalis longicornis)* and the tropical cattle tick (*Boophilus microplus)*. Potential climate change impacts on buffalo fly and sheep blowfly have also been inferred (Sutherst *et al.* 1996). Climate scenarios in New Zealand and Australia have indicated increased incidence of epidemics of animal diseases as vectors spread and extension of cattle tick which is directly related to changes in both temperature and rainfall (Sutherst, 1995). Incessant rains during 2010 monsoon resulted in severe Blue Tongue disease outbreak and huge mortality in sheep in coastal districts of Tamil Nadu, Karnataka and Andhra Pradesh due to heavy breeding of the vector, *Culicoides* sp. Heavy rains and high humidity during 2013 monsoon in Telangana resulted in Foot Rot disease epidemic in sheep (Venkateswarlu *et al*., 2011). Temperature and humidity variations could have a significant effect on helminth infections. Thus, in general, climate change-related temperature increase will have adverse impacts on the animal production system.

Innovative Adoption Options for Smart-Animal Agriculture

Adaptation helps in reducing vulnerability of people and ecosystems to climatic changes, where as mitigation reduces the magnitude of climate change impact in the long term. Animal keepers, especially resource poor farmers have a key role to play in promoting and sustaining a low-carbon rural path through proper agricultural technology and management systems. Mitigation measures could include technical and management options in order to reduce GHG emissions from animal agriculture, accompanied by the integration of it into broader environmental services. It is important to remember that the capacity of local communities to adapt to climate change and mitigate its impacts will also depend on their socio-economic and environmental conditions, and on the resources they have available.

Adaptation strategies augment tolerance of farm animals to extreme weather conditions, ability to survive, grow and reproduce in conditions of deprived nutrition, high incidence of parasites and diseases. Adaptation includes all activities like producing more feed and fodder, providing shelter, immunizing against endemic diseases, maintenance of hygiene etc., that help people and ecosystems reduce their vulnerability to the adverse impacts of climate change and minimize the costs of climate change associated natural disasters. There is no one-size-fits-all solution for adaptation, measures need to be tailored to specific contexts, such as ecological and socioeconomic patterns, and to geographical location and traditional practices. The first and foremost adaptation strategy that help in reducing the vulnerability of animal agriculture production systems include enhancing feed and fodder base both at household and community level. The following are some adaptation strategies help in reducing vulnerability of animal agriculture to climatic changes

a) Enhancing feed and fodder base both at household and community level: Climate change and climate variability eventually lead to either loss or low production of feeds, fodders and crop residues. Hence, the following strategies can be adapted to increase the feed availability inorder to sustain the productivity from animal agriculture.

Fodder production from arable lands: Non availability of arable land has been severely affecting the area under fodder cultivation. As a result, the green fodder availability both qualitatively and quantitatively is much lower than requirement and leading to many nutritional deficiencies ranging from energy, protein to micronutrients like minerals and finally lowered production from farm animals. Hence, each farmer should at least allocate 10% of their land for fodder production. Alternatively, farmers should be encouraged to cultivate winter crops like Lucerne, Horse gram etc., as second crop under rainfed conditions, where as fodder maize/ sorghum with little irrigation facilities. Further, on the bunds of all SWC structures, seeding of fodder varieties like Stylo and Cencherus must be encouraged. This helps in not only providing fodder but also stabilizes the bunds.

***Intensive irrigated fodder production systems*:** High yielding perennial (hybrid Napier varieties like CO-3, CO-4, APBN-1 etc.,) and multicut fodders varieties (MP Chari, SSG etc.,) could be choice of fodder crops under this system as it efficiently utilizes limited land resources and other agricultural inputs for getting maximum forage per unit area. It can be done where ever water is available and transported to deficit areas.

Intensive rainfed fodder production systems: Growing of two or more annual fodder crops as sole crops in mixed strands of legume (Stylo or cow pea or hedge Lucerne etc) and cereal fodder crops like sorghum, ragi in rainy season followed by berseem or Lucerne etc., in rabi season in order to increase nutritious forage production round the year

Short duration fodder production from tank beds: Due to silt deposition, tank beds are highly fertile and retain adequate moisture in the soil profile for cultivation of short season fodder crops like sorghum and maize fodder.

Integrated fodder production systems: Fodder crops like *Stylo hamata* and *Cenchrus ciliaris* can be sown in the inter spaces between the tree rows in orchards or plantations as hortipastoral and silvopastoral systems for fodder production.

Fodder production systems through alley cropping: Alley cropping is a system in which food/fodder crops are grown in alleys formed by hedgerows of trees or shrubs (*Leucaena leucocephala, Gliricidia, Calliandra, Sesbania etc.*). The essential feature of the system is that hedgerows are cut back at planting and kept pruned during cropping to prevent shading and to reduce competition with food crops. The main objective of alley cropping is to get green and palatable fodder from hedgerows in the dry season and produce reasonable quantum of grain and stover in the alleys during the rainy/cropping season. This calls for cutting back (lopping) of hedge rows during the dry season. A welcome feature of alley cropping is its ability to produce green fodder even in years of severe drought.

Perennial non-conventional fodder production systems: Perennial deep rooted top feed fodder trees and bushes such as *Prosopis cineraria, Hardwickia binata, Albizia* species, *Zizyphus numularia, Colospermum mopane, Leucaena leucocephala, Azadirachta indica, Ailanthus excelsa, Acacia nilotica* trees and modified plants of cactus are highly drought tolerant and produce top fodder. Sowing of inter spaces of tree rows with drought tolerant grasses such as *Cenchrus ciliaris, Cenchrus setigerus* and *Lasirius sindicus etc*., further enhance forage production from these systems.

Use of unconventional resources as feed: The available waste products form food industries like palm press fibre, fruit pulp waste, vegetable waste, brewers' grain waste and all the cakes after expelling oil etc., should be used as feed to meet the nutritional requirements of different species of farm animals.

Fodder production systems in homesteads: *Azolla*, a blue green algae which has more than 25 % CP and a doublinf time of 5-7 days can be grown in pits at backyards depending on the number of milch animals owned by the farmer. Azolla yield is much more than the perennial fodder varieties like APBN-1/CO-3 etc and is around 1000 MT per ha at the rate of 300 gm./sq.m/day even after taking into account of unused space between two beds. It is more nutritious than the leguminous fodder crops like lucerne, cowpea, berseem etc and can be fed to cattle, buffalo, sheep, goat and also poultry after mixing with concentrate mixture at the ratio of 1:1

Hydroponic fodder production systems: By this method, fodder can be produced in large quantities within 8 days from seed. These include barley, oats, lucerne and maize. Growing grass fodder systems hydroponically is now becoming popular in drought prone areas. Hydroponic fodder production however requires large investment in the form of a commercial greenhouse, continuous supply of water and power. The state governments must encourage entrepreneurs to take up this activity in chronically drought prone areas. The department itself can establish some units to start with.

Year-round forage production systems: Cultivation of a combination of suitable perennial and annual forages for year round nutritious fodder supply for the farm animals using limited water resources. It consists of growing annual leguminous fodders like cowpea or horse gram etc inter-planted with perennial fodders like Co-3, CO-4, APBN-1 varieties of hybrid Napier in kharif and intercropping of the grasses with berseem, Lucerne, etc., during rabi season.

Fodder production through contingency plan: During early season drought, short to medium duration cultivated fodder crops like sorghum (Pusa Chari Hybrid-106 (HC-106), CSH 14, CSH 23 (SPH-1290), CSV 17 etc) or Bajra (CO 8, TNSC 1, APFB 2, Avika Bajra Chari (AVKB 19)etc.,) or Maize (African tall, APFM 8 etc.,) which are ready for cutting by 50-60 days and can be sown immediately after rains under rainfed conditions in arable lands during kharif season. If a normal rain takes place in later part of the year, rabi crops like Berseem (Wardan, UPB 110, etc varieties), Lucerne (CO-1, LLC 3, RL 88, etc.) can be grown as second crop with the available moisture during winter. In waste lands fodder varieties like Bundel Anjan 3, CO1 (Neela Kalu Kattai), *Stylosanthes scabra,* etc. can be sown for fodder production (Prasad *et al.*, 2012). Under irrigated conditions, Intensive forage sequences recommended for Southern region may be followed

- Sorghum + Cowpea (3 cuts) – Maize + Cowpea – Maize + Cowpea.
- Hybrid Napier or Guinea or Setaria grass inter-planted with Lucerne (8-9 cuts) or Hybrid Napier + Subabul / Sesbania (9-11 cuts/year).
- Sudan grass + Cowpea (3 cuts) – M.P. Chari + Cowpea (three cuts).
- Para grass + Centro (*Centrosema pubescens*) (9-11 cuts/year).

In case of mid season drought, suitable fodder crops of short to long duration as mentioned above may be sown in kharif under rainfed conditions. Mid season drought affects the growth of the fodder crop. Once rains are received in later part of the season the crop revives and immediate fertilization help in speedy recovery. If sufficient moisture is available, rabi crops like Berseem (Wardan, UPB 110, etc. varieties), Lucerne (CO 1, LLC 3, RL 88, etc.) can be grown during winter. In waste lands fodder varieties like Bundel Anjan 3, CO-1 (Neela Kalu Kattai), *Stylosanthes scabra* etc., can be sown for fodder production.

As late season drought affects seed setting, normal short duration fodder crops may be sown. Avoid multicut fodder varieties under rainfed conditions. All the available fodder must be harvested before drying out to preserve nutritive quality. Depending on availability of moisture, rabi fodder crops especially low water requiring varieties of lucerne may be planted. Normal intensive fodder systems may be followed under irrigated conditions.

In case of complete or major failure of grain crops in Kharif, contingency strategies for ensuring fodder supplies include re-sowing with short to medium duration fodder varieties of millets, pulses or forage crops such as:

- Sorghum – varieties / hybrids CSV-17 and CSH 14 in red soils; CSH 16, CSH 18 and CSH 21 in black soils
- Bajra - short duration varieties like Rajko, JB, PSB-2, GHB-526, HHB-67, ICMH-356, Shraddha, GK-1004 or medium duration varieties like GHB-558, Proagro-9443 and for late assured rain fall areas in light to medium soils of Marathwada region varieties like AHB-251
- Finger millet - medium duration varieties like GPU 28, PR 202, HR 911 and Pusa Composite 612, MP 480 for second fortnight of July to first fortnight of August; short duration varieties: GPU 26, GPU 45, GPU 48 and Indaf 9 for late sown conditions from second fortnight of August to 20 September
- Maize – African tall, APFM 8, PEHM-3 and FH-3077 which produce some grain and fodder
- Intercropping cowpea – Varieties Bundel Lobia-1, CO 5, CO (FC) 8, IFC 8401, UPC 8705, DFC 1 and UPC- 625 after 8 to 10 rows of finger millet
- Rabi fodder crops like berseem (Mescavi, Wardan, UPB 110), Lucerne (CO 1, LLC 3, RL 88) should be sown in arable lands and tank beds.
- Current fallows should be used for fodder production by sowing short duration varieties of sorghum or bajra or ragi or maize or cow pea in kharif season and or berseem or Lucerne in rabi season.
- In wastelands, grasses like *Cenchrus ciliaris, C. setigerus, Chloris gayana*, Panicum maximum, *Desmanthus virgatus, Stylosanthes scabra* can be taken up to increase forage production.
- In areas that receive north east monsoon rains, multi-cut fodder varieties of sorghum (CO 27, Pant Chari-5 (UPFS- 32), COFS- 29 or pearl millet (Co-8) or maize (African tall) are recommended
- In areas that receive summer rains, fodder crops like cowpea and maize are recommended

b) Integrated farming systems: The second most important in building the resilience of animal agriculture production systems is promoting integrated farming systems. Integrated farming system besides generating higher productivity, it also produces sufficient food, fruits, vegetables etc., to the farm families. Several IFS models like (A) Conventional cropping; (B) crop + poultry (20) + goat (4); (C) crop + poultry (20) + goat (4) + dairy (1); (D) crop + poultry (20) + goat (4) + sheep (6); and (E) crop + poultry (20) + goat (4) + sheep (6) + dairy (1) were studied. Among the models examined, model (E) recorded a maximum net income of Rs 52794/ha, with maximum employment generation (389 man days/ha/year) (Solaiappan *et al.*, 2007). Integrated farming system comprising enterprises viz. field and horticultural crops, poultry, fishery (0.20 ha) and apiary (5 bee hive boxes) in 0.6 ha area in Chintapalli of high altitude tribal zone of Andhra Pradesh recorded a net income of Rs.29,102 and B:C ratio of 1.83 with productivity of 14.40 (t ha-1) and 464 man days/ha/year over arable cropping returns (Rs.14500/ha) and B:C ratio (1.47) with less productivity (7.50 t ha-1) (Sekhar *et al.*, 2014). Integration of field crops (Rice) + poultry + fish + horticultural crop (banana) resulted in highest system productivity (14.90 t ha- 1) in terms of rice grain equivalent yields. Further, integration of different farm components i.e., crops + horticultural crops (fruits & vegetables) and farm animals along with vermi-composting as value addition practice has been found to have maximum gross and net returns with maximum net returns of Rs. 42,610 (51.7%) from farm animals, including vermin-compost (AICRP-IFS,2013). Inclusion of 10-20 synthetic poultry breeds like Giriraja/Vanaraja/Gramapriya/Rajasree etc., at backyard with available food grain wastes/ grain byproducts (broken rice/ rice bran etc.) from the cropping system will also provide additional income through sale of eggs and chicken. All these types of systems are suitable for the scarce rainfall zone where the rainfall is 500-750 mm.

Crop-animal integrated systems are recommended for the areas having some irrigation facilities and or receiving above 1000mm rain fall with high yielding graded Murrah buffaloes and crossbred cows and crops. These areas generally produce surplus crop residues besides allocation of some cultivated land for fodder crops and purchase of feed supplements. In these systems inclusion of 10-20 synthetic poultry breeds like Giriraja/Vanaraja/Gramapriya/Rajasree etc., at backyard will further boost the income of the farmers. Crop, animal, poultry, fishery integrated farming system are mostly suitable for high rainfall areas, where paddy is cultivated both in *Kharif* and *Rabi* seasons. Cows and or buffaloes are maintained at backyard with crop residues and supplements. Fish is reared in farm ponds and poultry is maintained in cages over the pond with grain and bran supplementation. The droppings of poultry serve as feed for the fish in the pond.

Silvo-pastoral systems are efficient integrated land use management systems of agricultural crops, tree fodder species and or farm animals simultaneously on the same unit of land which results in an increase of overall production. Inter spaces between fodder trees species (*Leucaena leucocephala*) are utilized for cultivation of grasses and grass legume mixtures (*Cenchrus ciliaris* and *Stylosanthes hamata* or *scabra*), which provides a two tier grazing under *in situ*. This type of systems provides Rs.25000-30000 income per ha (Ramana, *et al*., 2000) and helps in reclamation of soil in waste lands and are more suitable for rearing small ruminants (10-12 animals/ha) in degraded waste lands under dryland conditions in Scarce rainfall zone. Horti-pastoral systems, the inter tree spaces in the mango/lemon/sweet orange orchards are utilized for cultivation of grasses and grass legume mixtures (*Cenchrus ciliaris* and *Stylosanthes hamata* or *scabra*) along with one side boundary plantation of fodder trees species (*Leucaena leucocephala*). Cultivated fodder and weeds serve as feed for the animals. Integration of lambs provide Rs.4000-5000 additional income per ha through sale of animals, control weeds by grazing/ browsing and also improve soil fertility through faeces and urine (Ramana, 2008 and Ramana *et al*., 2011)

c) Efficient utilization of available feed resources: Feeding and nutrition are the major constraints to intensification of animal agriculture under climate change scenario. Animal production within the mixed farming systems is predominantly dependent on the efficiency of use of the available coarse crop residues and grazing resources. The level of efficiency will dictate to a very large extent improvement per animal performance and increased productivity from different animal species. The limited purchasing power forces farmers to manage animals by carefully adjusting the resources and production factors of their farms. Hence, capacity development of the stakeholders about feeds and their efficient use should be imparted in the form of action learning exercise along with providing chaff cutters in each village, so that the stake holder would practically see how the chopped fodder is going to benefit not only to the animal but also to the farmer. Soaking of chopped crop residues with 2 per cent molasses/ jaggery solution or 1 percent each salt and molasses/ jaggery solution for 1-2 hrs and feeding enhances digestibility by 10-15%. Further, establishing complete feed mills at mandal level would help in efficient use of all the available resources including cotton stalks for feeding animals especially during lean season and or drought. In order to meet the nutrient requirement of huge animal population in India, it is essential to improve the fodder quality of cereal and legume food crops through crop improvement programs. The resulting new varieties would provide good yields of both human food, animal feed and mulching material. One should make greater use of crop residues as animal feeds, which will make better use of water resources by spreading the "cost" of the water used for growing crops across the grain and animal feed components.

d) Modifications in managemental practices: This includes feeding breeding and shelter management for different species of animals. Changes in animal management practices could include: (i) mixed crop animal farming systems, diversification, intensification and/or integration of fodder production (silvopastoral systems) and/or orchards (hortipastoral systems), alley cropping (ii) changing land use and irrigation; (iii) conservation of CPRs; (iv) modifying grazing practices (rotational grazing and or restricted grazing); (v) introducing especially during lean period, such as stall-fed systems through cut and carry fodder production; (vi) better feeding management through conventional and unconventional feed resources (vii) providing proper shelter and adequate wholesome water throughout the year (viii) identification and promotion of local breeds that have adapted to local climatic stress and feed sources; (ix) improvement of local animals through cross-breeding with heat and disease tolerant breeds and (x) synchronization of oestrus based on the availability of feed resources and favourable climatic conditions, (xi) supplementation of micro minerals and vitamins especially during lean season.

As most of the animal agriculture depends on common property resources (CPRs) for nearly about 60 per cent of daily feed requirement, proper management play a crucial role in sustainable production. As there is no control over the number of animals allowed to be grazed, causing severe damage on the regrowth of no. of herbaceous species in grazing lands. Thus causing severe impact not only on herbage availability from CPRs but also on the productivity of grazing animals. CPRs need to be reseeded with high producing legume and non-legume fodder varieties at every 2-3 years intervals as a community activity. Further, grazing restriction till the fodder grows to a proper stage as community decision would improve the carrying capacity of CPRs. Introducing grass species, legumes and tree fodder species into grazing lands and application of nutrients during the monsoon would enhance the availability of feed and fodder from CPRs and also enhances carbon sequestration in soils.

Further, use of sprinkler and or rain gun irrigation systems for fodder production, cleaning of animals with recycled water, high yielding and low water required multi-cut fodder varieties, improving CPRs productivity through re-sowing and weeding etc., would increase animal agriculture water productivity.

e) Eradication, containment and surveillance of diseases: Diseases results in i) reducing the animal population through death or culling; ii) reducing productivity of animals; iii) creating market shocks when demand falls and supply contracts in response; and iv) disrupting international trade in animal products. They decrease the productivity of animals by causing death or reducing the efficiency with which they convert feed into meat, milk and eggs. Hence, participatory disease surveillance, early forewarning, traceability and emergency systems would help in containment of disease epidemics. Ring vaccination (5km radius) and restriction of animal movement prevents spreading of contagious diseases in rural areas in the event of any outbreak.

f) Housing management: It is very essential especially under extreme weather conditions. The type of housing needed will depend on the climate, topography and the production system. Cost of construction, ease of cleaning, proper ventilation and drainage, hygiene maintenance and adequate lighting are important aspects to be considered in designing a shelter for animals. The basic requirement of good animal housing is that it should alter or modify the environment for the benefit of animals and also protect them from predation and theft. Animal housing should buffer the animal from climate extremes to reduce stress allowing optimal animal performance in terms of growth, health and reproduction. A proper housing should meet animal requirements and serve a producer's needs at the lowest possible cost.

g) Breeding management: Majority of the local breeds are having better adaptive capacity to harsh living conditions. Hence, one should (i) identify and strengthen local breeds that have adapted to local climatic stress and feed sources; (ii) improvement of local animals through cross-breeding with heat and disease tolerant breeds and (iii) synchronization of oestrus in farm animals based on the availability of feed resources and favourable climatic conditions

h) Technology development: Working towards a better understanding of the impacts of climate change on animal agriculture production systems, developing tolerant breeds, fodder varieties and specific vaccines, improving animal health and enhancing water and soil management would support adaptation measures in the long term.

i) Capacity building of animal keepers: It is essential to improve the capacity of animal producers and herders to understand and deal with climate change impacts in the long run. Capacity building training programmes on agroecology specific technologies and practices for the production and conservation of fodder and also better health management practices improves the supply of animal feed, reduces malnutrition and mortality in farm animals.

j) Institutional and policy support: Introducing subsidies for fodder production, establishment of complete feed mills, insurance of animals, income diversification practices and establishing farm animals early warning systems and other forecasting and crisis-preparedness systems could strengthen adaptation efforts.

Conclusions

Animal agriculture has greater role in economic development of the country and helping to lift people out of poverty and build sustainable livelihoods. It is essential to incorporate specific and regionally-sensitive policies and measures to ensure that food production is both humane and sustainable considering the positive role of farm animals in achieving sustainable agriculture. Use investment in research and development to promote and support low GHG emission and integrated animal agricultural systems to ensure sustainable climate resilient livelihoods for rural communities.

References

Naresh Kumar, S., Anil Kumar Singh, Aggarwal, P.K., Rao, V.U.M. and Venkateswarlu, B. (2012). Climate Change and Indian Agriculture: Impact, Adaptation and Vulnerability – Salient Achievements from ICAR Network Project, IARI Publication, p. 32.

Prasad, Y.G., Venkateswarlu,B., Ravindra Chary, G., Srinivasa Rao, Ch., Rao, K.V., Ramana, D.B.V., Rao, V.U.M., Subba Reddy. G. and Singh, A.K.. (2012). Contingency Crop Planning for 100 Districts in Peninsular India. Central Research Institute for Dryland Agriculture, Hyderabad-500 059, India. 326p.

Ramana, D.B.V. 2008. Silvopastoral and hortipastoral models for small ruminant production. In: Alternate land use systems for resource conservation, emerging market needs and mitigation of climate change in rainfed regions, 16th January-5h February 2009, CRIDA, Hyderabad, 239-249pp.

Ramana, D.B.V., Pankaj, P.K., Nikhila, M., Rita Rani and Sudheer, D. (2013). Productivity and physiological responses of sheep exposed to heat stress. Journal of Agro-meteorology, 15 (Special Issue-I): 71-76.

Ramana, D.B.V., Rai, P., Solanki, K.R and Singh, U.P. (2000).Comparative performance of lambs and kids under silvopastoral system. In: Proc. III Biennial ANA conference, Hissar, pp47-48.

Ramana, D.B.V., Reddy, N.N and Rao, G.R. (2011). Hortipastoral systems for ram lamb production in rain fed areas. Annals of Biological Research, 2011, 2 (4) : 150-158.

Sekhar, D, K. Tejeswara Rao, N and Venugopala Rao (2014). Studies on Integrated Farming Systems For Tribal Areas of Eastern Ghats In Andhra Pradesh. Indian Journal of Applied Research 4(10)14-15.

Solaiappan, U., Subramanian, V and Maruthi Sankar, G.R. (2007). Selection of suitable integrated farming system model for rainfed semi-arid vertic inceptisols in Tamil Nadu. Indian Journal of Agronomy 52 (3): 194-197.

Srivastava A.K. (2010). Climate Change Impacts on Livestock and Dairy Sector: Issues and Strategies. In: National Symposium on Climate Change and Rainfed Agriculture, February 18-20, 2010, Indian Society of Dryland Agriculture, Central Research Institute for Dryland Agriculture, Hyderabad, India. Pp 127-135.

Sutherst, R.W. (1995). The potential advance of pest in natural ecosystems under climate change: implications for planning and management. In 'Impacts of climate change on ecosystems and species: terrestrial ecosystems'. (Eds. J. Pernetta, C. Leemans, D. Elder, S. Humphrey) IUCN, Gland, Switzerland, pp83-98.

Sutherst, R.W., Yonow, T., Chakraborty, S., O'Donnell, C. and White, N. (1996). A generic approach to defining impacts of climate change on pests, weeds and diseases in Australia.

In 'Greenhouse: Coping with climate change'. (Eds. W.J. Bouma, G.I. Pearman and M.R. Manning.) pp. 281-307. (CSIRO: Melbourne.) 169-172.

Thornton P. and Herrero M., (2008). Climate Change, Vulnerability, and Livestock Keepers: Challenges for Poverty Alleviation. In Livestock and Global Climate Change conference proceeding, May 2008, Tunisia.

Upadhyay R.C., Ashok Kumar, Ashutosh, S.V. Singh and Avtar Singh (2007). Impact of climate change on milk production of crossbred cows. 4 th Congress of Federation of Indian Physiological Societies (FIPS), January 11-13, 2007, DIPAS, DRDO, Delhi.

Venkateswarlu, B., Singh, A.K., Prasad, Y.G., Ravindhra Chary, G., Srinivasa Rao, Ch., Rao, K.V., Ramana, D.B.V., Rao, V.U.M. (2011). District level Contingency Plans for Weather Aberrations in India. Central Research Institute for Dryland Agriculture, Natural Resource Management Division, ICAR, Hyderabad, India. 136p.

10

Climate Effect on High Yielding Animals: Adaptation and Mitigation Strategies

T K Mohanty and Mukesh Bhakat

Introduction

The animal husbandry and agriculture are the major resource of income for the farmers and directly affects the economic conditions of farmers. Livestock sector plays a vital role for livelihood food security in India. Buffalo plays an important role in dairy industry and contribute 49 per cent of total milk production in India (BAHS, 2017). Murrah is one of the best buffalo breeds in the world and also widely used for up-gradation of non-descript buffaloes in India. Murrah buffalo have poor thermoregulation mechanism compared to other domestic ruminants in tropical countries and are more prone to heat stress due to very less sweat glands, jet black colour and thin hairs on body surface (Das *et al*., 1999; Khongdee *et al*., 2013) which reduce the ability of cutaneous evaporation and largely responsible for its low productivity and consequently reducing the reproductive efficiency (Gudev *et al*., 2007). India continues to be the largest milk producer in the world since 1997. Milk production has increased from 165.4 million tonnes (2016-17) to 176.4 million tonnes (2017-18) (DAHD, GOI). The demand for milk is increasing at a very fast pace due to exponential population growth and the gap in the production and supply is increasing. Out of 300 million bovines, only 88 million are in milk leaving large unproductive animals including 84 million males (19th Livestock census, 2012). If the gap of production and supply is to be bridged, the large number of constraints that affect reproductive efficiency and productivity of dairy animals must be addressed effectively. One important factor that can contribute to the productivity of animals is qualitative production and quantitative adequacy of semen doses for A.I.

Climate change is one of the major challenges for livestock production systems in tropical countries. Sere *et al.* (2008) reported that heat stress has adverse effects on the productive, reproductive and health performances of dairy animals. Heat stress is a major factor contributing to the decline in fertility in lactating dairy cows (De Rensis and Scaramuzzi, 2003; Dash *et al.*, 2016). Several studies reported a 20 to 30% reduction in conception rate (Schuller *et al.*, 2014) and in pregnancy rate (Khan *et al.*, 2013) in hot climatic condition. However, an air temperature above 25-37°C exceeds heat gain than their lost from the body and it induces heat stress in a tropical climate (Vale, 2007; Kumar *et al.*, 2011). Increase in temperature of earth per decade by 0.2°C, global average surface temperature would be increased to 1.4-5.8°C by 2100. The major environmental factors affect livestock production system include temperature, relative humidity (RH), solar radiation, precipitation and wind speed (WS) (Hahn *et al.*, 2003). According to intergovernmental panel on climate change (IPCC, 2007), there is an increase in body surface temperature, rectal temperature (RT), respiration rate (RR) and pulse rate (PR) and decrease in feed intake, production and reproductive efficiency in hot climatic conditions. Strategies to ameliorate negative impact of heat stress on production and reproduction in dairy animals include improved housing and management intervention to reduce climatic impacts on livestock. Quality housing and various cooling systems such as use of fogger, mister and sprinkler with or without fan, force ventilation, feeding management, diet manipulation and use of assisted reproductive technology, has proven implication in management of reproduction efficiency. The cooling system is the most effective way to increase both milk production and reproduction in dairy animals during the summer season.

Environmental Factor and Animal Stress

Many environmental factors directly or indirectly affect on production performance of animals. Change in climatic condition directly affects the production and reproduction level of animal about 58.3% and 63.3%, respectively (Singh *et al.*, 2012). High environmental temperature leads to changes in the animal's body physiology such as rise body temperature (>102.5 °F), respiration rates (> 70-80/minute) and blood flow (Pereira *et al.*, 2008). The maintenance energy requirement may increase by 20-30% in animals under heat stress, which leads to reduced intake of feed and low energy level for productive functions such as milk production and increased loss of ions like sodium and potassium. This results shift in the acid-base balance and leads to metabolic alkalosis.

Temperature Humidity Index (THI) to Assessment of Heat Stress Level

Temperature-humidity index (THI) is the universal and most precise indicator of stress assessment using temperature and humidity data of the environment. Hot climatic conditions lead to decline production and nutrient intake of animal. The heat generated by metabolizing nutrient contributed to body temperature maintenance in a cold environment. However, in a hot climate, heat needs to be dissipated to maintain body temperature and normal physiological functions. Marai *et al.* (2008) reported that exposure of animals to hot climatic conditions lead to drastic changes in the biological functions which include decrease in feed intake and its utilization, disturbances in enzymatic activity, metabolism of water, protein, energy and mineral balances. THI is account for combined effects of environmental temperature and relative humidity and animal response. THI can be calculated with several formulae with varying accuracy in different climatic conditions. Milk yield decline by 0.2kg per unit increase in thermal humidity index (THI) when it exceeded 72 (Ravagnolo and Misztal, 2000). The production performance of cows is negatively correlated with temperature-humidity index (Shinde *et al.*, 1990; Mandal *et al.*, 2002). When the environmental temperature rises from the upper critical limit, the detrimental effects of heat stress on animals in terms of reduction in production of milk, changes in composition of milk and reduced reproductive performances are observed in cattle and buffaloes. Several studies report the classification of different zones based on THI values whether the animals are comfortable or susceptible to heat stress.

Temperature Humidity Index is negatively correlated to milk yield, an increase of THI value from 68 to 78 decreases DMI by 9.6% and milk production by 21% (Spiers *et al.*, 2004; Bouraoui *et al.*, 2002). Zimbelman *et al.* (2009) also reported a negative relationship between rectal temperature and milk yield of animal. Johnson *et al.* (1963) reported that decrease in milk yield by 4 lbs/d per cow for every 0.55 °C increase above the rectal temperature of 38.6 °C. Igono *et al.* (1985) reported that decrease in milk yield 0.7 kg/day per cow when temperature was increased to 0.6°F above the rectal temperature 102.4°F. Milk constituents are significantly affected by heat stress during summer season. Dairy breeds are more susceptible to heat stress than meat breeds, and higher milk-producing animal had increased metabolic heat production and this causes more susceptibility to heat stress as compared to low milk-producing animals (Das *et al.*, 2016). A decrease in protein constituent shows that reduction in casein, lactalbumin, IgG and IgA. Heat stress causes decline in dry matter intake and feed conversion efficiency which directly affects the body condition and resulting low milk yield (Wilson *et al.*, 1998).

Effect of Heat Stress on Animal Reproduction

The temperature humidity index is a common indicator of heat stress used in cattle and buffaloes for production performance in tropical and subtropical climatic conditions. The scale was originally established in 1960s after taking into account the combined effect of environmental temperature and relative humidity. Increased productivity of dairy buffaloes leads to a rise in metabolic heat production. Buffaloes show a poor thermal tolerance power due to presence of less number of sweat glands and underdeveloped thermoregulatory system (Marai and Haeeb, 2010). Hence they are unable to get rid of excess metabolic heat and susceptible to heat stress. Climate change has a great impact on the reproductive activity of cattle and buffaloes (Dash *et al.*, 2015). High temperature combined with high level of relative humidity has detrimental effect on reproduction of cattle in summer season. Heat stress had negative effect on reproductive traits of cattle and buffaloes which can be quantified through formulating temperature humidity index (THI). Conception rates of lactating dairy animals have been declined with increased THI more than 72-73 in cattle (Morton *et al.*, 2007; Schuller *et al.*, 2014) and 75 in buffalo (Dash, 2013). The release of ACTH from anterior pituitary, which stimulates the release of cortisol and glucocorticoids from adrenal cortex occurs during heat stress condition. The release of luteinizing hormone is also inhibited by glucocorticoids. The hyperprolactinaemia, as a result of thermal stress inhibits the secretion of both FSH and LH at hypophyseal level (Singh *et al.*, 2013).

In a study in NDRI herd by Upadhaya *et al.* (2012) in Buffaloes, it was reported that buffaloes have distinct estrus rhythm during normal ambient conditions and reproductive functions are influenced by change in Tmax and Tmin. Buffalo has a typical rhythmic pattern with one or two peaks during the year. The low reproductive activity was observed during summer when intensity of solar radiation is high and duration of sunshine is more. The frequency of estrus was pronounced during cool periods and buffaloes start exhibiting estrus after onset of monsoon. More buffaloes exhibit estrus when THI is near comfortable level (THI<70) in morning. The deviations from mean, in estrus during the years were observed mainly from July onwards reaching a peak in September or October months. A majority of buffaloes exhibited signs of estrus from September to November when ambient temperatures were low and comfortable. Climate change scenarios constructed for India revealed that maximum temperature (T max) rise of 4.54, 4.42, 3.07 and 4.38°C, respectively during Dec-Feb, Mar-May, Jun-Aug and Sep-Nov and very less change in precipitation is likely to increase uncomfortable days (THI>80). Increased number of heat stress days with THI>80 are likely to have a negative impact on estrus symptoms, duration and conception of buffaloes.

In a study by Dash *et al*., 2016 on buffaloes, the threshold THI for pregnancy rate was identified as 75 in Murrah buffaloes. McGowan *et al*. (1996) reported that there was significant ($P < 0.05$) decrease in first service pregnancy rate of dairy cattle and an increase in number of services per pregnancy with an increase in THI above 72 which corresponds to temperature 250 C and relative humidity 50% in Queensland. Morton *et al*. (2007) reported that when THI on the day of service is more than 72, it results in a decline in conception rate of dairy cows in Australia. Garcia-Ispierto *et al*. (2007) identified the negative association of conception rate of Holstein dairy cows with increase in THI in northeastern Spain where there was decrease in conception rate from 35-33% to 21-27% at THI values higher than 75. The water buffalo is the polyestrous animal and the sexual activities occur all-round year but it shows a distinct seasonal variation in the display of estrus and conception rate (Singh and Nanda, 1993). Ambient temperature and relative humidity showed a direct effect on breeding efficiency (Roy *et al*., 1968). Abayawansa *et al*., (2011) reported when there was maximum temperature during April and May; the percentage of buffalo exhibited postpartum estrus was lowest. There is a negative correlation ($r = -0.6$) between monthly postpartum oestrus incidence and mean maximum temperature. The lowest relative humidity during April and May was correlated with lowest incidence of postpartum oestrus. However, there is no research available on threshold THI affecting pregnancy rate of Murrah buffaloes.

Silent heat is one of the deleterious features to the reproductive performance in the buffaloes and it is due to poor oestrus expression and intensity of heat in buffaloes during summer as compared to winter (Madan and Prakash, 2007). Parmar and Mehta (1994) reported that weak symptoms of oestrous exhibited by buffaloes during summer are due to the lower synthesis and secretion of oestradiol-17 beta by the ovarian follicles. Heat stress causes hyperprolactinemia which results in the reduction of luteinizing hormone secretion and oestradiol production in anoestrus buffaloes (Palta *et al*., 1997) leading to ovarian inactivity. The survival of embryo in the uterus is impaired due to the deficiency of progesterone in the hot season (Bahga and Gangwar, 1988). This endocrine pattern may be partially responsible for the low sexual activities and low fertility in the buffalo in summer season. In summer due to non availability of good quality of feed and fodder results in poor reproductive performance of buffaloes. In buffaloes, undernutrition and high environmental temperature are the two major factors responsible for long anoestrous in Murrah buffaloes (Kaur and Arora, 1984).

Buffaloes in the subtropical region are exposed to heat stress due to high THI values from the month of April to September with a range 75.4 - 81.6. The climate in the months from October to March was favourable for Murrah buffaloes with a range of THI 56.7 - 73.2. In our study, months from October to March were considered as no heat stress zone with THI< 75 and April to September as heat stress zone with THI ≥75. Armstrong (1994) categorized THI values into four different classes as comfort zone (THI < 71), mild stress (72-79), moderate stress (80-89) and severe stress (> 90). Hisashi *et al.* (2011) in South-western Japan developed two periods in a year i.e. cool period from October to June with am THI under 75 and the hot period from July to September with am THI over 75. In northeastern Spain, Garcia-Ispierto *et al.* (2007) classified two periods in a year such as cool period including months from October to April with the range of THI 42.9 to 57.7 and warm period including months from May to September with the range of THI from 61.0 to 73.0.

Dash *et al.* (2015) classified a whole year into three different zones viz; non heat stress zone including months from October to March with mean THI 56.71 - 73.21, heat stress zone during the months of April to September with mean THI 75.39 - 81.60 for all the fertility traits. The months of May and June were identified as critical heat stress zone within the heat stress zone for days open and pregnancy rate while June, July and August were critical for the conception rate of Murrah buffaloes. Mc Dowell *et al.* (1976) developed three different classes of THI as comfortable (≤70), stressful (71-78) and extreme distress (> 78). Armstrong (1994) categorized THI values into five different classes as no stress with THI value < 72, mild stress (72-78), moderate stress (79-88), severe stress (89-98) and dead cows with THI > 98. The result revealed that May and June's months were the critical heat stress zone for days open and pregnancy rate, while June, July and August months were the critical heat stress zone for conception rate in Murrah buffaloes. The breeding value for fertility traits was influenced by THI. The EBV for fertility traits differed in three zones. In critical heat stress zone the EBV of days open was found around two days higher than non heat stress zone. A decline of -0.5% in EBV of pregnancy rate and -0.31 % in the EBV of conception rate was observed in critical heat stress zone as compared to non heat stress zone.

Effects of Heat Stress on Health of Dairy Animals

In a review by Sinha, *et al*, (2017), has described heat stress has direct and indirect effects on health performance of animal leading to changes in physiology, metabolism, hormonal and immune system. Increase in environmental temperature has a direct negative effect on voluntary feed intake and efficiency of feed utilization (Baile and Forbes, 1974). Lactating cows start to decline the feed intake at air temperature of 25-26°C and reduce more rapidly above 30°C in temperate climatic conditions and at 40°C it may decline by 40% in cattle, 8-10% in buffalo heifer and 22-35% in goat (Rhoads *et al*., 2013; Hamzaoui *et al*., 2012; Hooda *et al*., 2010). The increased environmental temperature may increase risk of metabolic disorders and health problems and change the basic physiological mechanisms resulting decreasing rumen motility and rumination (Nardone *et al*., 2010; Soriani *et al*., 2013). Heat stress changes metabolic patterns results in decreased thyroid activity and reduces metabolic heat production (Helal *et al*., 2010). Increase in ambient temperature causes increased incidence of in lameness in animals (Cook *et al*., 2007). This coincides with the change of climate as well the lameness prevalence is higher in hot climates as compared to cooler climates (Sanders *et al*., 2009). These climatic and seasonal effects are also correlated to mastitis in dairy animals (Dohoo and Meek, 1982; Elvinger *et al*., 1991).

Strategies to Ameliorate Heat Stress

To reduce heat stress is the multidisciplinary approach. It should include modification of microenvironment; nutritional management and genetic improvement are key components for sustainable livestock production under hot environmental conditions.

Modification of Micro Environment

Modification of microenvironment to improve heat dissipation mechanism to alleviate heat stress is one of the most important measures to be considered in hot environment. The most common approach to ameliorate heat stress is to modify environment near to cow way through provision of shade, evaporative cooling system by use of fogger, mister or sprinkler with fan or without fan (Atrian and Shahryar, 2012). Improve reproductive performance of cows using effective cooling systems that combine evaporative cooling with tunnel ventilation or cross ventilation (Kadokawa *et al*., 2012).

Nutritional Management

Lower DMI during hot weather reduces nutrients available for absorption, and absorbed nutrients are used less efficiently (West, 1999). Low-fibre, high fermentable carbohydrate diets lower dietary heat increment compared to high fibre diets. Although the metabolic energy of dairy buffaloes increases in a hot environment, heat stress depresses feed intake. Therefore, the strategies to increase the nutrient density includes feeding of high-quality forage, concentrates and use of supplemental bypass fat in the diet of animals. During hot climate, dietary fat content in feed is to be increased to enhanced milk production efficiency and yield. Supplementation of niacin supportive to reducing heat stress in cattle and supplementation with antioxidants during the heat stress period is added to improve fertility in buffaloes (El-Tarabany and Nasr, 2015). Both Vitamin C and Vitamin E have antioxidant properties. Antioxidant vitamins have proved to protect the biological membranes against the damage of ROS and the role of vitamin E as an inhibitor –"chain blocker"- of lipid peroxidation has been well established (Seyrek *et al*., 2004). Khan *et al.* (2011) in buffaloes has suggested that the amount of micronutrients needed for optimal immune function may exceed that which will prevent more classical signs of deficiency. In general, mineral deficiencies have been associated with altered metabolic profile leading to most peri-parturient disorders in buffaloes. Thus, such disorders could probably be prevented by addressing the basic aetiology through balanced feeding and mineral supplementation during advanced pregnancy and the early post-partum period, when the animals are highly prone to the stress of heavy nutrient demand and drain (Mandali *et al*., 2002). Besides general nutritional status, deficiency or imbalance with respect to specific nutrients has been found to have a drastic effect on various determinants of reproductive performance.Further, the impact of such disturbances on general health, including insidious subclinical diseases/ disorders, has been recognized as the most important but covert factor with deleterious consequences for reproductive performance. In this perspective, nutritional management for general and reproductive health has been considered to be of paramount importance (Mulligan *et al*., 2006). Greater understanding about the effects of better year-round nutrition, improved management and markers for logical breeding programmes are essential to curtailing incidence of the reproductive disorders that reduce buffalo fertility (Nanda *et al*., 2003).

Genetic Modification

The identification of heat tolerant animals within high producing breeds and they can be select genetically for crossbreeding programme to improve genetic variation and cooling capability (Kimothi and Ghosh, 2005). Cattle with lighter, thin skin, short hair and greater diameter of hair coat colour are more adapted to hot environments as compared to darker colours and long hair coats (Bernabucci *et al*., 2010). In a study by Dash *et al*., (2015) revealed that May and June's months were the critical heat stress zone for days open and pregnancy rate, while June, July and August months were the critical heat stress zone for conception rate in Murrah buffaloes. The breeding value for fertility traits was influenced by THI. The EBV for fertility traits differed in three zones. In critical heat stress zone the EBV of days open was found around two days higher than non heat stress zone. A decline of -0.5% in EBV of pregnancy rate and -0.31 % in the EBV of conception rate was observed in critical heat stress zone as compared to non heat stress zone.

Effect of Heat Stress on Bull

The most affordably adopted ART in livestock farms in India is artificial insemination. Seminal quality and bull fertility were decreased with high THI due to alteration of thermoregulation mechanisms of the testis and reduced their comfort (Kastelic and Brito, 2012; Santos *et al.*, 2014; Marai *et al.*, 2008). Various authors reported significant influences of season on semen quality parameters in breeding bull (Koivisto *et al.*, 2009; Teixeira *et al.*, 2011; Snoj *et al.*, 2013; Bhakat *et al.*, 2015). Alteration in semen quality parameters was due to the change in the seasonal expression of low-density lipid receptors on the surface of sperm cells which affect the exploitation of seminal plasma constituents in-between seasons (Argov *et al.*, 2007). The season has direct effect on animals through macro and microclimatic factors and indirect effect by affecting the feed and fodder quality (Mandal *et al.*, 2000). Winter season has indicated the high quality of semen as compared to other season (Schwab *et al.*, 1987). Infrared thermography is a noninvasive technique commonly used to assess the physiological and metabolic parameters and its impact on animal welfare (Kotrba *et al.*, 2007; Montanholi *et al.*, 2009). IRT has been reported to provide accurate information about scrotal surface temperature and helps in better understanding of the thermoregulatory mechanism. Testicular thermoregulation plays a vital role in the maintenance of spermatogenesis, and it must be 4 to 5 °C below the core body temperature to produce fertile sperm in bulls (Kastelic *et al.*, 1995). Elevated temperature causes mitochondrial dysfunction due to interference with the oxidative metabolism of glucose in sperms and the generation of reactive oxygen species (ROS). Scrotal insulation commonly used as experimental model to determine the effects of temperature on sperm production and semen quality and reported dramatically reduces sperm motility, live percentage and sperm concentration (Kastelic *et al.*, 2001; Burns *et al.*, 2010). Summer has been reported to be unfavourable condition for quality semen production in Murrah buffalo bulls (Bhakat *et al.*, 2015; Menegassi *et al.*, 2016) due to reduced feed intake, higher cortisol level and reduced release of reproductive hormone (Gilad *et al.*, 1993; Clark and Tibrook, 1992). The body and scrotal surface temperature can be measured by infrared thermography, which is a fast, accurate and non-invasive method, used in andrological evaluations and to understand the scrotal/testicular thermoregulation of bulls (Kastelic, 2014; Menegassi *et al.*, 2015). Therefore, present study was undertaken to assess the influence of season on in-vitro sperm functions, body and scrotal surface temperature of Murrah buffalo breeding bulls.

The demand for semen doses for A.I have been rapidly growing each year in the country due to limited availability of elite dairy bulls with detailed pedigree records. Semen production has increased from 22 million doses (1999-2000) to 115 million (2017-18) but the projected demand for semen doses is about 140 million doses (NDDB's internal compilation, 2017-18). It is required that semen production from elite bulls should be hastened and every ejaculate should be used in optimum way without affecting fertility of semen sample. Crossbred bulls donate nearly 20–30% poor quality semen owing to inherent problem, seasonal and prophylactic stresses rendering them unfit for use in AI, so it should be aimed to use good quality population of sperm in poor quality semen ejaculates without discarding the samples (Bhakat *et al*., 2016).

Conclusions

Climate change, in particular rising temperatures, can have both direct and indirect effects on animal production. Heat stress (caused by the inability of animals to dissipate environmental heat) can have a direct and detrimental effect on health, growth and reproduction. Changes in the nutritional environment (e.g. the availability of livestock feeds, and the quantity and quality of livestock pastures and forage crops) can have an indirect effect. These effects are expected to be most dramatic in temperate regions. Heat stress is a major economic issue in the dairy industry. It affects the production reproduction and health of animal through physiological changes. Environmental stress has adverse effects on the health status of dairy animals and decreases the milk production and reproductive performance of dairy cows resulting in huge economic losses. The most common method to reduce heat stress in dairy cows by provision of shades, sprinklers, ventilation and evaporative cooling will be suitable for adapting to climates changes. Environmental modifications and nutritional management are key elements to alleviate the impact of heat stress on animal's performance during the hot climate. Wallowing and sprinkling are the most effective methods to reduce heat stress in case of buffalo during summer season. Strategies to reduce negative impact of heat stress of animals using cooling system, ration manipulation, change in reproductive protocol, antioxidant, use of buffers, yeast and hormones will improve the economic status of dairy farmers. Climate change may affect zoonoses (diseases and infections which are naturally transmitted between vertebrate animals and man) in a number of ways.

References

Armstrong, D.V. (1994). Heat stress interactions with shade and cooling. J. Dairy Sci., 77: 2044-2050.

Atrian, P and Shahryar, H.A. 2012. Heat stress in dairy cows. Res. Zool., 2 (4): 31-37.

Baile, C.A and Forbes, J.M. 1974. Control of feed intake and regulation of energy balance in ruminants. Physiol. Rev., 54 (1): 160.

Bernabucci, U., Lacetera, N., Baumgard, L.H., Rhoads, R.P., Ronchi, B and Nardone, A. (2010) Metabolic and hormonal acclimation to heat stress in domesticated ruminants. J. Anim. Sci., 4(7): 1167-1183.

Bouraoui, R., Lahmar, M., Majdoub, A., Djemali, M. and Belyea, R. (2002). The relationship of temperature humidity index with milk production of dairy cows in a Mediterranean climate. Anim. Res., 51(6): 479-491.

Cook, N.B., Mentink, R.L., Bennett, T.B. and Burgi, K. (2007). The effect of heat stress and lameness on time budgets of lactating dairy cows. J. Dairy Sci., 90: 1674-1682.

Dash, S. (2013). Genetic evaluation of fertility traits in relation to heat stress in Murrah buffaloes. M.V.Sc. Thesis, ICAR-NDRI (Deemed University), Karnal, Haryana, India.

Dash, S., Chakravarty, A.K., Sah, V., Jamuna, V., Behera, R., Kashyap, N. and Deshmukh, B. (2015). Influence of temperature and humidity on pregnancy rate of Murrah buffaloes. Asian-Aust. J. Anim. Sci., 28(7): 943-950.

Dash, S., Chakravarty, A.K., Singh, A., Shivahre, P.R., Upadhyay, A., Sah, V. and Singh, K.M. (2015) Assessment of expected breeding values for fertility traits of Murrah buffaloes under subtropical climate. Vet. World, 8(3): 320-325.

Dash, S., Chakravarty, A.K., Singh, A., Upadhyay, A., Singh, M. and Yousuf, S. (2016) Effect of heat stress on reproductive performances of dairy cattle and buffaloes: A review. Vet. World. 9(3): 235-244.

De Rensis, F. and Scaramuzzi, R.J. (2003). Heat stress and seasonal effects on reproduction in the dairy cow-a review. Theriogenology., 60: 1139e51.

Dohoo, I. R. and Meek, A. H. (1982). Somatic cell counts in bovine milk. Can. Vet. J., 23: 119-125.

El-Tarabany, M.S. and El-Bayoumi, K.M. (2015). Reproductive performance of backcross Holstein x Brown Swiss and their Holstein contemporaries under subtropical environmental conditions. Theriogenology., 83: 444-448.

Elvinger, F., Hansen, P. J. and Natzke, R. P. (1991). Modulation of function of bovine polymorphonuclear leukocytes and lymphocytes by high temperature in vitro and in vivo. Am. J. Vet. Res., 52:1692-1698.

Hahn, G.L., Mader T.L. and Eigenberg, R.A. (2003). Perspective on development of thermal indices for animal studies and management. EAAP tech. series, 7: 31-44.

Hamzaoui, S., Salama, A.A.K., Caja, G., Albanell, E., Flores, C. and Such, X. (2012). Milk production losses in early lactating dairy goats under heat stress. J. Dairy Sci., 95(2): 672-673.

Helal, A., Hashem, A.L.S., Abdel-Fattah, M.S. and El-Shaer, H.M. (2010). Effect of heat stress on coat char-acteristics and physiological responses of Balady and Damascus goats in Sinai, Egypt. Am. Euresian J. Agric. Environ. Sci., 7(1): 60-69.

Hooda, O.K. and Singh, S. (2010). Effect of thermal stress on feed intake, plasma enzymes and blood bio-chemicals in buffalo heifers. Indian J. Anim. Nutr., 27(2): 122-127.

Igono, M. O., Steevens, B. J., Shanklin, M. D. and Johnson, H. D. (1985). Spray cooling effects on milk production, milk and rectal temperatures of cows during a moderate summer season. J. Dairy Sci., 68: 979-985.

Intergovernmental Panel on Climate Change (IPCC). (2007) Climate Change: Synthesis Report. Available from: http://www.ipcc.ch/pdf/assessment report/ar4/syr/ar4_syr_sym.pdt. Accessed on 28-11-2015.

Johnson, H. D., Ragsdale, A. C. Berry, I. L. and Shanklin, M. D. (1963). Temperature-humidity effects including influence of acclimation in feed and water consumption of Holstein cattle. Missouri Agr. Exp. St. Res. Bul. 846.

Kadokawa, H., Sakatani, M. and Hansen, P.J. (2012). Perspectives on improvement of reproduction in cattle during heat stress in a future Japan. Anim. Sci. J., 83(6): 439-445.

Khan, F.A., Prasad, S. and Gupta, H.P. (2013). Effect of heat stress on pregnancy rates of crossbred dairy cattle in Terai region of Uttarakhand, India. Asian Pac. J. Reprod., 2(4): 277-279

Kimothi S.P. and Ghosh C.P. (2005). Strategies for ameliorating heat stress in dairy animals. Dairy Year book. 371-377.

Kumar, B.V., Kumar, A. and Kataria, M. (2011). Effect of heat stress in tropical livestock and different strategies for its amelioration. J. Stress Physiol. Biochem., 7(1): 45-54.

Mandal, D.K., Rao, A.V.M.S., Singh, K. and Singh, S.P. (2002). Effects of macroclimatic factors on milk production in a Frieswal herd. Indian J Dairy Sci., 55(3):166–170.

Marai, I.F.M, El-Darawanya, A.A. Fadielc A. and Abdel-Hafezb, M.A.M. (2008). Reproductive performance traits as affected by heat stress and its alleviation in sheep. Tropical and Subtropical Agroecosystems, 8: 209 – 234.

McDowell, R.E. (1972). Improvement of livestock production in warm climates. Freeman, San Francisco, C.A., pg 711.

Morton, J.M., Tranter, W.P., Mayer, D.G. and Jonsson, N.N. (2007). Effect of environmental heat on conception rates in lactating dairy cows: Critical periods of exposure. J. Dairy Sci., 90: 2271-2278.

Nardone, A., Ronchi, B., Lacetera, N., Ranieri, M.S. and Bernabucci, U. (2010) Effect of climate changes on animal production and sustainability of livestock systems. Livest. Sci., 130(1-3): 57-69.

Pereira, A.M.F., Baccari Jr, F., Titto, E.A.L. and Almeida, J.A.A. (2008). Effect of thermal stress on physiological parameters, feed intake and plasma thyroid hormones concentration in Alentejana, Mertolenga, Frisian and Limousine cattle breeds. International Journal of Biochemistry. 52: 199-208.

Ravagnolo, O. and Misztal, I. (2000). Genetic component of heat stress in dairy cattle, parameter estimation. J. Dairy Sci., 83: 2126-2130.

Rhoads, M.L., Rhoads, R.P., Baale, M.J., Collier, R.J., Sanders, S.R., Weber, W.J., Croocker, B.A. and Baumgard, L.H. (2009). Effects of heat stress and plane of nutrition on lactating Holstein cows: I. Production, metabolism, and aspects of circulating somatotropin. J. Dairy Sci., 92(5): 1986-1997.

Rhoads, R.P., Baumgard, L.H., Suagee, J.K. and Sanders, S.R. (2013). Nutritional interventions to alleviate the negative con-sequences of heat stress. Adv. Nutr., 4(3): 267-276.

Sanders, A.H., Shearer, J.K. and De Vries, A. (2009) Seasonal incidence of lameness and risk factors associated with thin soles, white line disease, ulcers, and sole punctures in dairy cattle. J. Dairy Sci., 92(7): 3165-3174.

Schuller, L.K., Burfeind, O. and Heuwieser, W. (2014). Impact of heat stress on conception rate of dairy cows in the moderate climate considering different temperature humidity index thresholds, periods relative to breeding, and heat load indices. Theriogenology., 81: 1050-1057.

Sere, C, Zijpp, A.V., Persley, G. and Rege, E. (2008). Dynamics of livestock production system drives of changes and prospects of animal genetic resources. Anim. Genet. Resour. Inf., 42: 3-27.

Seyrek, K., Kargin Kiral, F. and Bildik, A. (2004). Chronic ethanol induced oxidative alterations in the rat tissues and protective effect of vitamin E. Ind. Vet. J., 81: 1102-1104.

Shinde, S, Taneja, V.K. and Singh, A. (1990). Association of climatic variables and oduction and reproduction traits in crossbreds. Indian Journal of Animal Sciences. 60(1): 81–85.

Singh, M., Chaudhary, B.K., Singh, J.K., Singh, A.K. and Maurya, P.K. (2013). Effect of thermal load on buffalo reproductive performanceduring summer season. Journal of Biological Sciences. 1(1):1-8.

Singh, S.K., Meena, H.R., Kolekar, D.V. and Singh, Y.P. (2012). Climate change impacts on livestock and adaptation strategies to sustain livestock Production. Journal of Veterinary Advances. 2(7): 407-412.

Soriani, N., Panella, G. and Calamari, L. (2013). Rumination time during the summer season and its relationships with metabolic conditions and milk production. J. Dairy Sci., 96(8): 5082-5094.

Spiers, D.E., Spain, J.N., Sampson, J.D. and Rhoads, R.P. (2004). Use of physiological parameters to predict milk yield and feed intake in heat-stressed dairy cows. J. Therm. Biol., 29(7-8): 759-764.

Upadhyay, R.C., Ashutosh and Singh, S.V. (2009). Impact of climate change on reproductive functions of cattle and buffalo. In: Aggarwal, P.K., editor. Global Climate Change and Indian Agriculture. ICAR, New Delhi. p107-110.

Vale, W.G. (2007). Effects of environment on buffalo reproduction. Ital. J. Anim. Sci., 6(2): 130-142.

West, J.W. (1999). Nutritional strategies for managing the heat-stressed dairy cow. American Society of Animal Science and American Dairy Science Association, 2: 21-35.

West, J.W. (2003). Effect of heat stress on production in dairy cattle. J. Dairy Sci., 86: 2131-2144.

Wheelock, J.B., Rhoads, R.P., Van Baale, M.J., Sanders, S.R. and Baumgard, L.H. (2010) Effect of heat stress on ener-getic metabolism in lactating Holstein cows. J. Dairy Sci., 93(2): 644-655.

Wilson, S J., Marion, R.S., Spain, J.K., Spiers, D.E., Keisler, D.H. and Lucy, M.C. (1998). Effect of controlled heat stress on ovarian function of dairy cattle. J. Dairy Sci., 1: 2124-2131.

Zimbelman, R.B., Muumba, J., Hernandez, L.H., Wheelock, J.B., Shwartz, G., O'Brien, M.D., Baumgard, L.H. and Collier, R.J. (2007). Effect of encapsulated niacin on resistance to acute thermal stress in lactating Holstein cows. J. Dairy Sci., 86: 231.

11

Heat Stress and Its Implications on Large Animals

Swagat Mohapatra and A K Kundu

Introduction

Stress represents the reaction of body to stimuli that disturb normal physiological equilibrium or homeostasis, often with detrimental effects (Khansari *et al.*, 1990). Domestic animals undergo various kinds of stress such as physical, nutritional, chemical, psychological and thermal stress (Marai *et al.*, 2007; Nardone *et al.*, 2010). Figure 1 summarises the different types of stressors and their effects on animals.

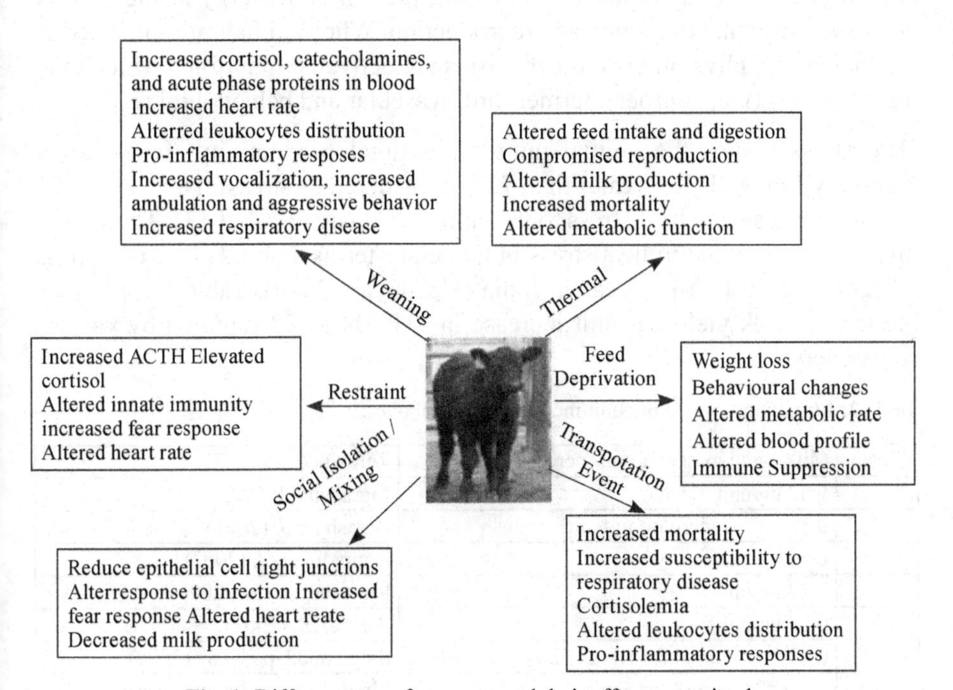

Fig. 1: Different type of stressors and their effects on animals

Among all, the thermal stress is most concerning now-a-days in the ever changing climatic scenario. Current climatic models indicated an increase in temperature of the earth by 0.2°C per decade and predicted that the increase in global average surface temperature would be between 1.8 °C to 4.0 °C by 2100 (Intergovernmental Panel on Climate Change, 2007). The thermoneutral zone (TNZ) of dairy animals in India ranges from 16°C to 25°C (Das *et al.*, 2016). The ambient temperature range that is considered comfortable for large animals is represented in table 1.

Table 1: Range of ambient temperature considered comfortable for large animals

Animals	Temperature Range (°C)	Reported by
Holstein cow	5-21	Worstels and Brody, 1953
Jersey and Brown Swiss cows	5-24	Johnson, 1965
Crossbred cow	15-25	Goswami and Narain, 1962
Buffaloes	13-24	Singh and Upadhyay, 2008

Temperature determines metabolic rates, heart rates and other important factors within the bodies of animals, so an extreme temperature change can easily distress the animal body. The effect of high temperature is further aggravated when heat stress is accompanied by high ambient humidity. Thermal stress redistributes the body resources including protein and energy at the cost of decreased growth, production and reproduction. When animals are subjected to thermal stress, physiological mechanisms are activated manifolds which may include endocrine, neuroendocrine, cardiovascular and behavioural responses.

Heat stress makes the animal unfit for optimal productivity. Temperature Humidity Index (THI) greater than 72 is considered as stressful and THI > 78 is considered severe heat stress to the animal (Ganaie *et al.*, 2013). The loss of milk production due to heat stress in monetary terms amounts to a whopping Rs. 26616.2 million per year in India (Upadhyay, 2010). Table 2 represents the loss in milk yield per unit increase in THI above 72 reported by various researchers.

Table 2: Loss in milk yield per unit increase in THI above 72

Sl. No.	Milk Yield loss/THI unit increase	Reference
1.	0.32 kg/unit	Ingraham (1979)
2	0.38 kg /°C increase of temp.	Barash *et al.* (2001)
3.	0.41 kg/unit	Bouraoui *et al.* (2002)
4.	0.2 kg/unit	Ravagnolo *et al.* (2000)
5.	0.88 kg/unit	West *et al.* (2003)
6.	0.43lit./d/animal for Karan Fries	Singh and Upadhyay (2008)
7.	0.16lit./d/animal for Sahiwal cows	Singh and Upadhyay (2008)

8.	(a) 0.36 kg/unit (High yielders, 32 kg/day)	Herbut and Angrecka (2012)
	(b) 0.28 kg/unit (Avg. yielders, 21 kg/day)	Herbut and Angrecka (2012)
	(c) 0.18 kg/unit (Low yielders, 12 kg/day	Herbut and Angrecka (2012)
9.	0.26 kg/day	Brown – Brandl *et al.* (2003)

The crossbred / exotic cattle are more prone to the heat stress losses as compared to indigenous cattle. (Belsare and Pandey, 2008). It has been established that only 35% of the reduction in milk yield during high ambient temperature is due to decreased feed intake remaining 65% reduction is due to direct physiological effect of heat stress (Rhoads *et al.,* 2009). Even in artificially cooled shed heat stress can decline milk production by 10-15% (Baumgard *et al.*, 2006). Animals try to neutralise the effect of heat stress on them by making certain physiological, behavioural and metabolic responses. However, the responses of animals can nullify the effect of heat stress up to a certain extent beyond which the animal's productivity is bound to get hampered. Singh *et al.,* 2018 noted some visible and invisible signs of heat stress in dairy animals.

Visible signs of heat stress in dairy animals

- Decreased activity.
- Animals seek shade and/or wind.
- Refusal to lie down.
- Increased respiratory rate (open mouth panting).
- Open mouth and laboured breathing.
- Sweating and excessive drooling.
- Reduced food intake.
- Agitation and restlessness.
- Thirst is increased. Drinking water intake increases markedly (5 times temperate climates.
- Increased urination (with heavy electrolyte loss).
- Crowding over the water troughs.
- Excessive salivation.
- Drop in milk yield
- Loss of milk quality - fat and protein content declines.
- Loss of body weight.

- The incidence of milk fever increases.
- Metritis is more widespread.
- Uterine prolapse is more common.
- Mammary gland infections increase.
- There are increased uterine infections.
- Udder oedema is more severe.
- Laminitis is more frequent.
- Keto-acidosis is a recurring problem.
- Insemination success rate falls i.e., Fertility is lowered.
- Increased somatic cell counts and risk of Mastitis.
- Embryo mortality increases.
- Calves are often premature and small.
- Growing animals have markedly reduced weight gains.
- Inability to move.
- Collapse, convulsions & coma.

Invisible signs of heat stress in Dairy Cattle

- Ruminal pH is typically lower in heat stressed cattle. Rates of gut and ruminal motility are reduced, thus slowing passage of feed through the digestive tract.
- Increased peripheral blood flow.
- Some indigestibility of feed.
- The huge water flux resulting from increased water consumption also causes heavy loss of electrolytes.
- Alter the production of reproductive hormones essential for pregnancy. Changes the balance of developing follicles in the ovary.
- Embryonic development is affected.
- Bicarbonate is lost.
- Stress hormones appear in the blood.
- Gene function is disturbed.
- Keto-acidosis is a recurring problem.
- Heat shock proteins are activated to shut down metabolic reactions and to protect heat-sensitive tissues.
- Responses to concurrent diseases or pathogens decline rapidly.

- Resources being diverted to unproductive efforts by the animal to restore balance (homeostasis).
- All production is stopped due to loss of homeostasis.
- The animal has done all that it can do to stop deep body temperature from rising and assistance is needed to restore, and retain, electrolyte balance to the optimum range.
- Heat stress is acidogenic and on the acid side of pH 6.8 there is probably widespread cell damage.

Effect of heat stress on cardiac health

A healthy heart is a pre-requisite for optimum milk production in cattle. Around 500 litres of blood must pass through the mammary gland for production of 1 litre of milk (Reece, 2015). Total udder blood volume for lactating cows is about 8% of total body blood volume, while for a non-lactating cow it is about 7.4%. Olsson *et al.,* (2000) reported that the mean cardiac outputs in goats were 35 and 28 % greater during pregnancy and lactation, respectively versus the dry period. Stroke volume of cattle heart averages around 600 ml, heart rate averaged about 60 beats per minute and cardiac output varies between 28 to 52 litres per minute (Doyle *et al,* 1960). Heat stress induced increase in certain serum cardiac biomarkers viz plasma aspartate aminotransferase, creatine kinase and lactate dehydrogenase activities while causing severe heart damage, which was characterized by granular and vacuolar degeneration, nuclear shrinkage and even myocardium fragmentation in cardiac muscle fibres of chicken (Zhang *et al.,* 2016). Previous studies reported that heat stress could cause severe damage to myocardial cells in rats, accompanied by an increase in apoptotic cells (Islam *et al.,* 2013). Studies on humans have suggested an increase in serum myocardial enzymes during heat stroke indicating myocardial damage (Alzeer *et al*., 1997). Similar impact of heat stress on myocardium of chickens was reported by Wu *et al*., (2016). Damage to myocardial cells leads to their replacement by fibrous tissue which reduces ventricular compliance and affect ventricular filling. So, any injury to myocardial cells of cattle during heat stress will definitely affect the pumping effectiveness of heart which would subsequently affect cardiac output and milk production which needs further investigation.

Strategies for ameliorating the effects of heat stress

The objective of mitigating the effect of heat stress in large animals could be achieved by creating awareness about the derogatory effects of heat stress on animals' health and productivity among farmers, veterinarians and stake holders. The strategies to ameliorate the effects of heat stress are based on shelter management, nutritional management and genetic modification.

Trees are effective blockers of solar radiation and the evaporation of moisture from leaf surface cools the surrounding air. So, the trees are an excellent natural shade under which animals can be restrained during hot weather conditions. Portable shades offer some advantages in their ability to be moved to a new area. Portable shade cloth, as well as light roofing material, may be used on the temporary shades. Mechanical cooling systems, high-rise ventilated type shed equipped with air circulators, pressurized showers for animal washing, in-line fan type spray nozzle for cleaning and cooling, high-rise head to head type animal housing, artificial shade with natural ventilation mechanism are certain modifications that can be used effective under shelter management to alleviate heat stress. Nutritional management includes feeding the animals in cooler hours of the day (morning and evening), providing green fodder, mixing of dry fodder with green fodder, grazing during morning and evening hours etc. Soaking of concentrate in equal amount of water for 20-30 minutes helps in better utilization of nutrients and reduces dustiness in concentrates, lush green fodder feeding also reduces the stress. Surplus green as silage and hay should be conserved. Dietary electrolyte balance should be maintained by adding essential body salts and electrolytes to drinking water and feed. Extra minerals should be supplemented in feed due to low dry matter intake and reduced availability of minerals to animals. Ample fresh, clean, cool and good quality drinking water should be available to animals as water requirements increases with temperature and intake may increase as much as 50% during periods of heat stress. Genetic modification includes the identification of heat tolerant animals within high producing breeds and they can be select genetically for crossbreeding programme to improve genetic variation and cooling capability (Kimothi and Ghosh, 2005). Cattle with lighter, thin skin, short hair and greater diameter of hair coat colour are more adapted to hot environments as compared to darker colours and long hair coats (Bernabucci *et al.*, 2010).

Climate change is a definite reality. Indian agriculture is likely to suffer losses due to heat, erratic weather and decreased irrigation availability. Adaptation strategies can help in minimizing the negative impacts which research and policy support. Cost of adaptation and mitigation are unknown but likely to be high but the cost of inaction could be even higher.

References

Belsare VP and Pandey V. 2008. Management of heat stress in dairy cattle and buffaloes for optimum productivity. Journal of Agrometeorology 10 (special issue): 365-368.

Ganaie AH, Shanker G, Bumla NA, Hasura RS, Mir NA, Wani SA and Dudhatra G B. 2013. Biochemical and physiological changes during thermal stress in bovines. Journal of Veterinary Science and Technology. 4: 126.

Khansari DN, Murgo AJ and Faith RE. 1990. Effect of stress on the immune system. Immunology Today, 11: 170-175.

Marai IFM, El Darawany AA, Fadiel A, Abdel Hafez MAM. 2007. Physiological traits as affected by heat stress in sheep a review. Small Ruminant Research. 71: 1- 12.

Nardone A, Ronchi B, Lacetera N, Ranieri MS and Bernabucci U. 2010. Effects of climate changes on animal production and sustainability of livestock systems. Livestock Science.130: 57- 69.

Rhoads ML, Rhoads RP, VanBaale MJ, Collier RJ and Sanders SR. 2009. Effects of heat stress and plane of nutrition on lactating Holstein cows: I. Production, metabolism and aspects of circulating somatotropin. Journal of Dairy Science, 92: 1986- 1997.

Singh N, Bharti P, Baranwal A, Kumar V and Pandey A. 2018. Ameliorative measures to counteract heat Stress in dairy animals of hot sub-humid eco-region. International Journal of Livestock Research, 8(3): 296-309.

12

Strategic Feeding Practices for Sustainable Livestock Production in Rain-fed Areas

R K Swain, S K Mishra and Kamdev Sethy

Introduction

Livestock plays an important role in the sustainable livelihood of poor people of rain-fed agro-ecosystem, because of inherent risk involved in the crop farming due to uncertainty of rainfall and occurrence of recurrent droughts. They provide income and increased economic stability, and often the most important "cash crops" in small-scale mixed farming systems. Livestock sector in India is highly livelihood intensive and most rural households own livestock of one species or the other and earn supplementary incomes from them. Livestock holding is less iniquitous than land holding and income from livestock is more equitably distributed. The sector contributes 25.6 per cent of the value of output at current prices of total value of output in agriculture, fishing & forestry sector and in GDP, it was 4.11% in 2012-13 (19th livestock census, 2012) therefore, development of the livestock sector is the critical pathway to rural prosperity

As per the 19th livestock census, the total livestock population consisting of Cattle, Buffalo, Sheep, Goat, Pig, Horses & Ponies, Mules, Donkeys, Camels, Mithun and Yak in the country is 512.05 million numbers in 2012. The total Bovine population (Cattle, Buffalo, Mithun and Yak) is 299.9 million numbers in 2012 which shows a decline of 1.57% over previous census. In percentage basis distribution of livestock were: 37.28% were cattle, 21.23% buffaloes, 12.71% sheep, 26.40% goats and 2.01% pigs. The corresponding figures as per the 18th Livestock Census were 37.58%, 19.89%, 13.50%, 26.53% and 2.10%. The number of milch animals (in-milk and dry) in cows and buffaloes has increased from 111.09 million to 118.59 million, an increase of 6.75%. The number of animals in milk in cows and buffaloes has increased from 77.04 million to 80.52 million showing a growth of 4.51%. The Female Cattle

(Cows) Population has increased by 6.52% over the previous census (2007) and the total number of female cattle in 2012 is 122.9 million numbers. The Female Buffalo population has increased by 7.99% over the previous census and the total number of female buffalo is 92.5 million numbers in 2012. The buffalo population has increased from 105.3 million to 108.7 million showing a growth of 3.19%. The total sheep in the country is 65.06 million numbers in 2012, declined by about 9.07% over census 2007. The Goat population has declined by 3.82% over the previous census and the total Goat in the country is 135.17 million numbers in 2012. The total poultry population in the country has increased by 12.39% over the previous census and the total poultry in the country is 729.2 million numbers in 2012. The total number of animals in milk in the country is 116.77 Million numbers.

Livestock production is the livelihood for 70% population in rural areas in India and its contribution to GDP is increasing year by year. Due to rapid urbanization, increase in purchasing power of the people and change in food habit, the demand for livestock products is increasing. The number of livestock population as per the 2050 report of Indian grass land and forage research institute (IGFRI) is presented in Table 1.

Table 1: Projected livestock population estimates$_{\#}$ (million Adult Cattle Unit, ACU)

Year	Cattle	Buffalo	Sheep	Goat	Equine	Camel	Total
2020	129.1	95.31	5.03	10.32	0.63	0.43	240.8
2030	133.6	106.8	5.39	11.18	0.54	0.29	257.9
2040	136.6	115.0	5.76	11.99	0.40	0.20	270.1
2050	139.6	127.1	6.13	13.19	0.29	0.12	286.5

Estimates based on past Livestock censuses published by the Directorate of Economic & Statistics and Department of Animal Husbandry & Dairying.

$_{\#}$Category-*wise population data was multiplied with standard body weight to get total weight while conversion to ACU (1 ACU = 350 kg).*

Source: 2050 report of IGFRI, Jhansi

In India, income from livestock production accounts for 15-40 % of total farm household earnings (World Bank, 1999). Milk production contributes on an average 27 percent of the household income; its contribution varies from about 19 percent in the case of large farmers to about 53 percent in the landless category in India. Apart from the monetary benefits provided by milch animals, the role of small ruminants like goats and sheep is very important, as they serve as a lifeline during drought years by providing income and sustenance. The small ruminants contribute 15 to 27 % of family income of smallholders and provide gainful employment of 180 to 330 man-days per annum depending on the size of the flock. It has also been shown that irrespective of flock size, women and children contribute to labour force to the extent of about 90 % (Deoghare 1997).

Thus, livestock provide income and increased economic stability, and often the most important "cash crops" in small-scale mixed farming systems. Small stock acts as cash buffer and large ruminants as a capital reserve. These assets can be realized at any time, adding security to the production systems. In India, about 75 % of rural households are small and marginal farmers, who own 56 % of the large ruminants and 62 % of the small ruminants (World Bank, 1999). In most of the drylands and hill regions, livestock farming is a major player, as more than 70 % of family income is derived from livestock (GOI, 2002). The Government of India has recognized the livestock development as an important tool for poverty alleviation. However, most of the technical service oriented livestock projects have had little impact on livelihood of the poor and adoption of technology by the resource poor has been low. The two major reasons indicated for this failure are lack of poverty focus and institutional framework being incompatible with the systems and conditions in which poor farmers stay. The institutions implementing the projects are probably unable to select appropriate interventions and approaches to reach poor farmers in an effective manner. In order to solve these problem, approaches that guarantee effective linkages among researchers, NGOs, extension workers, decision-makers and farmers, are required and some researchers now believe that "participatory approaches are mandatory" for the development of livestock technologies particularly forage production.

Livestock sustainability

The high economic demand for milk and meat in developed countries has led to the establishment of capital intensive systems that require high animal productivity, high levels of concentrate (grain) feeding and highly mechanized or automated infrastructure. The production technologies developed over the years have successfully met these requirements. The problems currently faced by these systems arise basically from economic pressures and revolve around capital costs, input costs and subsidies. Feeds comprise, by far, the largest component of input cost. The efficiency of feed utilisation in terms of herd or flock output, is the primary consideration in technology development.

Requirement vis a vis availability of feed resources in India

Detailed study on the requirement and availability of feed resources is not carried out to the full extent. As per the NIANP report, the availability of nutrients, feed and fodder availability and their projected requirement is presented in Table 2. As per the NIANP report for 2050, the shortage of crop residues, greens and concentrates works out to be 32, 25 and 47% respectively and it has projected the deficit of same will be 21.3, 40.0 and 38.1 per cent respectively by the year 2025

Table 2: Availability and requirement of feed sources

Feed resource	Availability (MT)	Requirement (MT)
Dry Fodder	433	550
Green Fodder	600	1000
Concentrates	65	105

Source: 2050 report of NIANP, Bangalore

The demand and supply estimates of dry and green forages (million tonnes) are presented in Table 3 and the projected requirement, availability and deficit of CP and TDN (million tonnes) is presented in Table 4. The projected feed requirement of animals as per the report of NIANP, 2050 is presented in Table 5. The data reveals that the there is wide gap between demand and supply and consistent and sincere effort is imperative to bridge the gap to sustain the increasing production of milk, meat and egg.

Table 3: Demand and supply estimates of dry and green forages (million tonnes)

Year	Demand		Supply		Deficit		Deficit as%	
	Dry	Green	Dry	Green	Dry	Green	Dry	Green
2010	508.9	816.8	453.2	525.5	55.72	291.3	10.95	35.66
2020	530.5	851.3	467.6	590.4	62.85	260.9	11.85	30.65
2030	568.1	911.6	500.0	687.4	68.07	224.2	11.98	24.59
2040	594.9	954.8	524.4	761.7	70.57	193.0	11.86	20.22
2050	631.0	1012.7	547.7	826.0	83.27	186.6	13.20	18.43

Source: 2050 report of IGFRI, Jhansi

Table 4: Projected requirement, availability and deficit of CP and TDN (million tonnes)

Year	Requirement		Availability		% Deficit	
	CP	TDN	CP	TDN	CP	TDN
2010	60.04	347.8	42.95	271.3	28.47	21.99
2020	62.58	362.5	47.18	290.5	24.60	19.87
2030	67.01	388.2	53.09	320.2	20.78	17.52
2040	70.19	406.6	57.61	342.8	17.92	15.69
2050	74.44	431.2	61.92	364.5	16.81	15.47

Source: 2050 report of IGFRI, Jhansi

Table 5: Feed requirement (on dry matter basis) for livestock in 2050

Category	Feed requirement for milch animals (mt)			Feed requirement for non- milch animals (m.t)			Total Feed requirement (m.t)		
	Conc	Dry	Green	Conc	Dry	Green	Conc	Dry	Green
Crossbred cattle	97	46	63.9	11.9	95	107	108.9	141	170.9
Indigenous cattle	9.5	14.3	7.9	0	24.2	18	9.5	38.5	25.9
Buffaloes	53.1	40.7	41	8.3	66.4	74.7	61.4	107.1	115.7
Total	159.6	101	112.8	20.2	185.6	199.7	179.8	286.6	312.5

Source: 2050 report of NIANP, Bangalore

Efficient feeding strategies

The feed is the main driver of livestock production. It accounts for up to 70% of the total cost of livestock operation. The poor or unbalanced feeding adversely affects the productivity, health, behaviour and welfare of animals. In addition, this also diverts a substantial portion of feed carbon and nitrogen to wasteful products in the form of GHG. Globally, the production, processing and transport of feed account for 45% of the GHG emission from the livestock sector. The area dedicated to feed-crop production represents 33% of total arable land and the grazing land constitutes 30% of the terrestrial land. Feed production is highly resource demanding. Approximately 33% and 6% of the grains produced are used for livestock feeding and bioethanol production, respectively. The food-feed-fuel competition is one of the complex challenges, and so are the ongoing climate change, land degradation and water shortages that need addressing for sustainable intensification of livestock production and for realization of sustainable food production and consumption systems. By 2050 the world population is expected to be 9.6 billion, which will require 60–70% more meat and milk than consumed today. Most of this increase will be from developing countries, which already face many food security challenges. Livestock use about 60% of the biomass used for food production. Ruminant livestock consume 78% of this biomass used and convert crop residues and by-products into edible products. Most of the dry matter consumed by livestock is composed of grass (39%) and other non-humanly edible materials such as crop residues (26%) and agricultural by-products (bran, oilseed cakes etc., 8%). Technologies are available that enhance digestibility of crop residues and by-products and also increase nutrient availability from them to animals i.e. increase feed conversion efficiency. Given that feed is by far the dominating physical flow, in energy terms, increase in feed conversion efficiency enhances overall resource use efficiency.

Use of unconventional feed resources

Various unconventional feed resources used in animals for sustainable production has been presented in Table 6.

Table 6: Uconventional feed resources, toxic factors and detoxification methods

Byproduct	Availability (Million tones)	Toxic factor	Detoxification	Nutritive value	Feeding value
Silk cotton seed meal	-	Cyclo-propenoid acid, fibre & tannins	Lysine and methionine supplemen-tation	20% CP, 1.30 Mcal ME/kg	Poultry 10%, may be fed to rumi-nants
Kosum cake	0.30	HCN	-	15%DCP, 79% TDN	Cross bred calves (35% in conc. Mix)
Linseed meal	0.57	Linamarin, linatin	-	6%DCP, 50% TDN	Milch cow (57% in conc. Mix)
Mango seed kernel	11.00	Tannins	-	6% DCP, 50% TDN	Milch cows (10% in conc mix.)
Niger seed cake	0.10	--	-	32% CP, 49% TDN	Crossbred calves (57% in conc. Mix)
Palm kernel meal	1.97	High fibre	-	20% CP, 1.90 Mcal ME/kg	Commonly fed to ruminants
Panewar seed	0.03	Crys-ophanic acid, Tannins	Fresh water washing / boiling	16% DCP, 66% TDN	Milch cows (15% in conc., mix.)
Safflower meal	1.49	Oxalates, Phytate in hull fraction	-	40% CP, 2Mcal ME/ kg	Poultry (15%) with lysine & mineral supplementation
Spent Anna-to seeds	0.30	Tannic acid	-	8% DCP, 67% TDN	Crossbred calves (20% in conc. Mix)
Sunflower meal	3.28	Chloro-genic acid, Tannins	Supple-menting with Methyl donors (Me-thionine & Choline)	36% CP, 2.10 Mcal ME/kg	Poultry (5%)
Ruber seed cake	0.15	HCN	Foasting, Toasting, Wa-ter soaking	18% DCP, 54% TDN	Milch cow (25% in conc. Mix.)
Tamarind seed	8.70	Tannins	-	1.3% DCP, 67% TDN	Calf starter (25%)

Densified straw based total mixed rations

Crop residues such as straws and stovers are valued feed resources in developing countries and they form 50–60% of the ruminant diets. The efficient technologies are now available for collection of straws from the crop fields. The collected straws can be used to form total mixed ration (TMR) by mixing with, for example, locally available oil seed cakes, urea, molasses, vitamin and mineral mixtures; followed by compacting to form blocks or pellets using hydraulic press. The TMR based on densified straw based blocks or pellets supply balanced feeds to animals and increase their productivity, resulting in profitability increase for farmers. Farmers find this technology attractive because use of a complete ration in the form of blocks/pellets decreases the feeding time. This is of particular importance for women because they are the main care takers of animals in developing countries. Time saved in feeding empowers women because they can use this time in other productive purposes. This technology can also be effective in disaster management and emergency situations that arise due to natural calamities, for example floods, droughts and man-made conflicts. Feed banks could be set up to overcome the problem of feeding animals during these natural calamities, which are common in the tropics.

Chopping of roughages

Simple technologies, such as chopping forages, increase animal productivity and reduce forage waste. Both intake and rumen digestion of chopped forages are higher than the un-chopped forages. Animals use a considerable amount of energy in chewing forages and chaffing saves this energy and diverts it for productive purposes. It decreases feed wastage and indirectly increases feed use efficiency

Urea molasses multi-nutrient blocks

The crop residues and grazing pastures during dry periods are deficient in nitrogen, energy and minerals. Urea molasses block supplementation enhances the supply of nitrogen, minerals and vitamins to rumen microbes which increases the nutrient supply to the ruminants from fibrous feed stuffs, thus enhancing their efficiency of utilization. Further, feeding crop residue with urea molasses blocks has resulted in the increased cost: benefit ratio ranging from 1:2 to 1:5 depending on the cost of feed and sale price of milk. In the recent years, use of urea molasses or multi-nutrient blocks during prolong winter period or severe drought has gained much attention. These blocks could also be used as a carrier for anthelminthic and tannin-neutralizing agents such as polyethylene glycol

Urea-ammoniation or CaO treatment of straw

Treatment of straws with 4-5% urea at 50-60% moisture level, followed by anaerobic fermentation for 15-20 days (depending on ambient temperature) increases digestibility by 10-15% units. This leads to higher productivity. Instead of urea, calcium oxide treatment can also be used to treat straws and stovers. This strategy has been effective in replacing substantial portion of grain in cattle diet thus reducing the food-feed competition and also enhancing the profit compared with the untreated corn based ration

Ensiling and converting waste to resources

Silage-making is also an attractive approach for reducing wastage of forages whose availability is high in rainy seasons. In some months of the year availability of vegetable and fruit wastes is also high which can also be converted into valuable resources through silage making. These resources can be used for feeding during the dry season when availability of feed is low.

Fodder Development

Fodder production by the farmers is possible only if they can spare land for that purpose if they have access to extra irrigation. In a predominantly subsistence farming system, supporting very low producing non -descript animals, the question of large scale promotion activities for fodder development does not arise. Therefore focus should be on selected areas with more number of cross bred population for promotional activities on fodder cultivation. Alternative fodder plants that are adapted to harsh climatic conditions and can be grown on lands unsuitable for cropping have significant potential for reducing the feed gap for livestock, improving animal health and contributing to food security and diversifying the income source of resource-poor farmers (Nefzaoui *et al.*, 2014). As supplementation legumes in the diet reduces the cost of animal production and in dry areas where livestock are an integral part of the production system, forage legumes have increased crop yield and animal production (Ates *et al.*, 2015).

Improvement of pasture land

Mmanagement of native pasture, introduction of suitable legume species etc., to increase the nutritive value of these resources could be effective followed by controlled grazing. Back yard hedge rows of fodder trees, social forestry and farm forestry initiatives have already been taken up by the government. In watershed areas special programmers are under way for creating alternate fodder resource for livestock especially for goats. Management of CPR is a more complex issue and needs to be addressed with great care. Infact, natural and environmental factors such as drought and loss of grazing lands have been identified as the most important causes of a decrease in number of livestock (Akter *et al.*, 2008).

Feeding of bypass nutrients to ruminants

The term "Bypass Nutrient" refers to that fraction of the nutrients which gets fermented in the rumen to a comparatively low degree. It then becomes available at the lower part of the gastro-intestinal tract in the intact form for subsequent digestion and absorption. They increase the protein supply to the animals (Preston and Leng, 1987). The protected protein by-passes the rumen and provides additional essential amino acids for absorption at the small intestine. These slowly degradable proteins also have a function to provide the rumen microbes with a steady supply of nutrients, rather than with sudden bursts from easily soluble nutrients. These concepts were introduced in the early eighties, primarily to replace the conventional digestible crude protein system which has many limitations. They describe the protein quality of a feedstuff for ruminants, and the protein requirements based on rumen degradable protein (RDP) and undegradable dietary protein (UDP). Subsequently, the term has been extended to other nutrients like carbohydrates and fats that could also escape rumen fermentation partially, to be digested in and absorbed from the small intestines. The new approach envisages minimization of ruminal fermentation losses, and better utilisation of the nutrients after their digestion and absorption from the small intestines. The purpose of feeding "bypass" protein is that a large proportion of the protein is available directly at the lower part of gastro-intestinal tract, where it is digested and then absorbed as amino acids for utilization at tissue level. Feeding of "bypass" starch reduces excess production of lactic acid in the rumen which would otherwise result in low rumen pH (acidosis), thereby affecting fibre digestion. Feeding of "bypass" fat (protected fat) is done primarily to avoid ruminal hydrolysis of bio-hydrogenation of unsaturated fatty acids and increasing energy density of feeds. The fats are thus digested mostly in the small intestines and absorbed as unsaturated fatty acids without affecting the fermentation of fibrous feeds in the rumen. Feeding bypass protein increases feed intake and as a consequence promotes milk production. Due to bypass protein supplementation, the availability of required aminoacid at intestine increased and supplies precursors of milk production. Supplementation of calcium salt of fatty acids in the diet had a positive effect on reproductive performance of dairy cows, which is further dependent up on the specific fatty acids profile of the Ca salt. Feeding calcium salt of fatty acids increases pregnancy rate and reduces open days (Sklan *et al.*, 1991). Garg and Mehta (1998) observed that the BSC of the cows improved due to bypass fat feeding indicating reduction in weight loss in the first quarter and helped gaining substantially after 90 days of feeding. Hammon *et al.* (2008)

observed that milk and lactose yields were higher in rumen protected fat fed animals than control. Adding a by-pass protein supplement will further improve the efficiency of utilization of the feed resources but will also allow animals to maintain feed intake at high environmental temperatures and humidity.

Inclusion of feed additives

Feed additives are ingredients or a mixture of ingredients, added to the basic feed mix, usually in small quantity, to fulfill a specific function, nutritive or non-nutritive. Some of these feed additives and their utility in the ration of livestock is mention below:

a) ***Enzymes:*** among the various enzymes, the use of fibrolytic/ NSP degrading enzymes is gaining popularity as it enhances digestion there by allows use of wide range of fibre/ NSP rich ingredients without compromising animal performance.

b) ***Probiotics:*** Probiotics are live microbial feed supplement, which beneficially affects the host animals by improving microbial balance. The probiotics preparation is generally composed of organism of lactobacilli and/ or streptococcus species. These organism have antagonistic effect on undesirable microorganism through production of antimicrobial compounds.

c) ***Presbiotics:*** Prebiotics are feed ingredients/ no digestible food that may affect the host by selectively stimulating the growth and activity of a limited number of bacteria in the colon. Fruocto-oligosaccharides are normally used as prebiotics. It is most widely used for prevention of diarrhea associated with intestinal infection.

d) ***Symbiotics:*** Combining prebiotics and probiotics in what has been called a symbiotic which could beneficially affects the host by improving survival of health promoting bacteria.

e) ***Acidifier:*** Organic acid usage in animal feeds to improve growth performance has gained importance in recent period. Organic acid like citric, formic, fumaric and propionic have pronounced effect in weaning pig diet and to greater extent in poultry.

f) In addition to these, anti-oxidants, emulsifier, pellet binders and mould inhibitors are used for prevention of oxidation of fat, binding of pellets during pelleted feed preparation and prevents development of mycotoxins in the feed, respectively.

Minerals in the ration

Minerals are virtually involved in every biochemical reaction or metabolic process in animals in regulating osmotic pressure, maintaining acid base balance, in control of membrane permeability and tissue irritability. Availability of minerals to animals in appropriate quantities is a major factor determining the health and productivity of animals. Mineral imbalances in soil and forages have long been responsible for low production and reproduction among ruminants. The extent and pattern of mineral deficiencies and excesses in plants vary in different agro-climatic conditions. The area specific mineral mixture prepared in different states for their agroclimatic zones should be included in the ration to avoid the deficiency of minerals. Mohapatra *et al.* (2012) reported that supplementation of area specific mineral mixture increased the reproductive and productive parameters of cows. Supplementation of minerals in nano forms enhances its availability and performance of animals.

Spineless cactus

Cultivation of spineless cactus (*Opuntia ficus-indica*) in degraded and marginal lands produces feed in water deficient conditions and also offers possibilities for carbon sequestration and land reclamation. It does not like saline and water logging conditions, but thrives in dry conditions, uneven rainfall and poor soils. It has the potential of not only decreasing the carbon dioxide levels in the atmosphere through the gas exchange pattern, termed as, Crassulacean Acid Metabolism (CAM) but also controlling of soil erosion by providing cover and enhancing afforestation. A biomass yield of 180 tonnes per hectare per year has been recorded in Brazil, and under mixed cropping systems with barley a yield varying from 25 to 100 tonnes has been obtained in Tunisia. The cactus cladodes are low in nitrogen but high in energy and water. A diet containing 60% cactus pods, 20% chopped hay and 20% protein rich concentrate mixture can support a cow yielding 25 litres milk per day (Dubeux *et al.*, 2013).

Insect meals

Food waste can also be used as a substrate for rearing insects such as black soldier fly larvae, maggot meal, mealworm larvae, which contain approximately 50% crude protein with good amino acid composition and can replace 50% of the conventional feed resources such as soya meal and fish meal in the diets of poultry and fish. These approaches convert 'disposal problems into opportunities for development (Makkar *et al.*, 2014)

Distillers grains

These are co-products of the bioethanol industry. Cereals such as maize, wheat, sorghum and barley are fermented to bioethanol. The mass of the dried distillers grains recovered after distillation of bioethanol are approximately one-third of the cereal mass taken for bioethanol production. Global yearly production of distillers grains is approximately 48 million tonnes. It is extensively used as livestock feed. The use of dried distillers grains with soluble (DDGS) as a substitute for the higher priced corn and soya bean in the diets of cattle, pigs, poultry and fish has been recorded, although the optimum levels of inclusion are still being determined. Distillers grains with soluble or with added protein (HP-DDGS) can be fed to pigs at all stages of the production chain.

Leaf meals and protein isolate

Moringa oleifera is a very fast growing plant. Moringa if grown as a fodder plant, contains on an average 16-17% crude protein while the leaf meal (without twigs and stems) contains 25-26% crude protein. The quality of moringa protein, in terms of essential amino acid composition and protein digestibility is very high - as good as soya meal. Under intensive cultivation conditions, moringa protein yield per hectare could be almost 5-times higher than that of soybean. Moringa leaf meal is also good source of sugars, vitamins and antioxidants. Moringa leaf meal could be a good replacer of soya meal in monogastric diets, while the twigs and soft stems could be fed to ruminants

Marine byproducts

Algae co-products and seaweeds could be good sources of protein and minerals. Brown seaweeds have been more studied and are more exploited than other algae types for their use in animal feeding because of their large size and ease of harvesting. Brown algae are of lesser nutritional value than red and green algae, due to their lower protein content (up to approx. 14%) however brown algae contain a number of bioactive compounds. Red seaweeds are rich in crude protein (up to 50%) and green seaweeds also contain good protein content (up to 30%). Seaweeds contain a number of complex carbohydrates and polysaccharides. Brown algae contain alginates, sulphated fucose-containing polymers and laminarin; red algae contain agars, carrageenans, xylans, sulphated galactans and porphyrans; and green algae contain xylans and sulphated galactans. These could be used as prebiotic for enhancing production and health status of both monogastric and ruminant livestock.

Conclusion

Management of natural feed resources for feeding the livestock is to be given importance considering the production potential of the high yielding animals. Due to scarcity of traditional feed ingredients and pasture land, judicious use of these sources and inclusion of unconventional feed resources may address the problem of deficit of feed resources to some extent. Improvement of poor quality roughages is to be emphasized for its maximum utilization.

References

Akter, S., J. Farrington, P. Deshingkar and A. Freeman 2008. Livestock, vulnerability, and poverty dynamics in India. ILRI Discussion Paper No. 10. London, UK and Nairobi, Kenya: Overseas Development Institute and International Livestock Research Institute.

Ates, S., Keles, G., Inal, F., Gunes, A. and Dhehibi, B. 2015. Performance of indigenous and exotic× indigenous sheep breeds fed different diets in spring and the efficiency of feeding system in crop– livestock farming. The Journal of Agricultural Science, 153(3) 554-569.

Deoghare, P. R. 1997 Sustainability of on-farm income and employment through livestock production in Mathura district of Uttar Pradesh. India Journal of Animal Sciences 67:916-919.

Dubeux, J.C.B., Dos Santos, M.V.F., de Mello, A.C.L. 2013. Forage potential of cacti on drylands. VIII International Congress on Cactus Pear and Cochineal 1067. 181-188.

Garg, M.R. and Mehta, A.K. 1998. Effect of feeding bypass fat on feed intake, milk production and body condition of Holstein Friesian cows. Indian Journal Animal Nutrition, 15:242-245.

GOI (Government of India). 2002. Report of the working group on Animal Husbandry and Dairying for the tenth five year plan (2002-2007). Working Group Sr. No. 42/2001. Planning Commission, Government of India, New Delhi, India.

Hammon, H.M., Metges, C.C., Junghans, P., Becker, F., Bellmann, O., Schneider, F. 2008. Metabolic changes and net portal flux in dairy cows fed a ration containing rumen-protected fat as compared to a control diet. Journal of dairy science. 91(1):208-217

Makkar, H.P.S., Tran, G., Heuzé, V.2014. State-of-the-art on use of insects as animal feed. Animal Feed Science and Technology, 197: 1-33

Mohapatra P, Swain RK, Mishra SK, Sahoo G and Rout KK. 2012. Effect of supplementation of area specific mineral mixture on reproductive performance of the cows, Indian Journal of Animal Science, 82 (12): 1558-1563.

Nefzaoui, A., Louhaichi, M. and Ben, S. H. 2014. Cactus as a tool to mitigate drought and to combat desertification. Journal of Arid Land Studies, 13: 121-124

Sklan, D., Moallem, U., Folman, Y. 1991. Effect of feeding calcium soaps of fatty acids on production and reproductive responses in high producing lactating cows. Journal of dairy science, 74(2):510-517.

Vision 2050.Indian Grassland and Fodder Research Institute (Indian Council of Agricultural Research) Gwalior Road, Jhansi - 284 003.

13

Nutritional Intervention in Changing Climatic Situation

Kamdev Sethy, R K Swain and S K Mishra

Introduction

Climate change is one of the biggest environmental threats to food production, water availability, forest biodiversity and livelihoods. Warming of climate system of the earth is a unanimously accepted reality and probably one of the most prominent challenges for scientists, development workers, policy makers and other relevant stakeholders regarding development and sustainability in international and national arena during past several years. Intergovernmental panel on climate change (IPPC) has described climate change as any anthropogenic or naturally occurring alteration in the climate over time. It is widely believed that developing countries such as India will be impacted more severely than developed countries. Climate change is emerging as a big threat to sustainable development of agriculture including animal husbandry and to the livelihood of people. Extreme events such as cold waves, heat waves, floods and high intensity single day rainfall events are on increasing trend during the last decade (Dikshit and Birthal, 2010). It is probable that by 2030 the impacts of climate change will be noticeable, although not dramatic and will be largely manifested through changes in pasture growth and quality, and greater inter-annual variability in pasture production. Livestock contribute to climate change by emitting methane through enteric fermentation. But they are also affected by climate change it directly and indirectly, hence affecting their economic and social contributions. Various studies suggest that the changes in pasture growth, composition and production will be dependent on the actual combination of CO_2, temperature and rainfall conditions. Livestock productions under both intensive and extensive systems are subjected to so many stresses, where they undergo different types of physiological adjustments to cope up with stressful conditions. Productions are drastically reduced during stress and require managemental interventions both in terms of optimum nutrition and health care (Soren, 2013).

Potential effects of climate change on livestock

To remain healthy and productive, livestock must regulate their body temperature within a relatively narrow range. The ambient temperature below or above the thermoneutral range creates stress conditions in animals. In India livestock are raised in mixed farming system where in animals depend on crop residues and byproducts for meeting their feed requirement. By adversely affecting the crop production and grazing resources, climatic shocks reduce supplies of feeds and fodders, and also their quality. High environmental temperature reduces productive and reproductive efficiency of livestock drastically. Adverse environments can increase the nutritional requirements of animals directly, or they may reduce the supply and quality of the feed. The total of heat produced in the course of digestion, excretion and metabolism of nutrients is called heat increment. Within a certain range of ambient temperature and besides unvarying feed and nutrient intake, the total heat production of the animal remains constant. This temperature range is called the thermoneutral zone (TNZ) and can be defined as the range of effective ambient temperature within which the heat from normal maintenance and productive functions of the animal is non-stress situations offsets the heat loss to the environment without requiring an increase in rate of metabolic heat production.

Thermoregulation is the ability of the animals to maintain their body temperature in cold or hot environments, consisting of behavioural, physiological and anatomical responses that affect energy metabolism. In homoiotherms it is achieved by physiological and behavioural adjustments which involve the musculature, skin, sensory capacities, hypothalamus and endocrine glands. Under thermal stress animals exhibit anorexia, body extension, gasping, languor lethargy, excessive drinking, bathing, decreased locomotor activities, group dispersion, and shade seeking. On the other hand, when animals are exposed to cold they show body flexure, huddling, hyperphagia, extra locomotor activities, depressed respiration and nest building. From a practical perspective higher temperatures are much more hazardous for growing/finishing and breeding animals than a cold environment. In general, livestock with high production potential are at greatest risk of heat stress, thereby requiring the most attention (Niaber and Hahn, 2007). Temperature-humidity index (THI) is used as an indicator of thermal climatic conditions. THI is determined by equation from the relative humidity and the air temperature and is calculated for a particular day according to the following formula (Kadzere *et al.,* 2002):

$$THI=0.72\ (W+D)+40.6$$

Where W - wet bulb temperature °C
D - dry bulb temperature °C

THI values of 70 or less are considered comfortable, 75-78 stressful, values greater than 78 cause extreme stress.

The thermal comfort zone for most animal ranges between 4 and 25°C and temperature exceeding 25°C will result in heat stress. It causes reduced feed intake, altered endocrine status, reduction in rumination and nutrient absorption, and increased maintenance requirements (Collier *et al.*, 2005) resulting in a net decrease in nutrient/energy availability for production. A reduction in energy intake in lactating cows during heat stress result in a negative energy balance (NEB). The NEB associated with the early postpartum period is coupled with increased risk of metabolic disorders and health problems, decreased milk yield and reduced reproductive performance (Baumgard *et al.*, 2006).

Animals respond to elevated temperature by reducing feed intake, increasing respiratory rate and water consumption, and by decreasing activity in an attempt to improve heat loss and minimize the heat generation in the body. In poultry, heat stress can be either acute or chronic. Chronic stress has deleterious effects on birds that are reared on open-sided houses and causes a reduction in feed consumption and increasing water intake. For every increase (1°C) in environmental temperature from 22 to 32°C, feed intake in broiler chicken was reduced by 3.6% (Ain Baziz *et al.*, 1996). High environmental temperature reduce milk yield in dairy cows especially in animals of high genetic merit. Johnson *et al.* (1962) demonstrated a linear reduction of dry matter intake (DMI) and milk yield when temperature humidity index (THI) exceeded 70. The reductions were -0.23 and -0.26 kg/day per unit of THI for DMI and milk yield, respectively

During extreme cold the outside temperature falls rapidly and substantial amount of dietary energy may be diverted from productive functions to the generation of body heat. Failure to produce sufficient heat can result in death. More often cold stress leads to the development of secondary changes and possibly disease. Cold-stressed animal will oxidize acetogenic substrate for heat production until the surplus acetogenic substrate is totally utilized, after which fat mobilization provides an extra source of metabolic fuel. The DMI of the animals in general is increased during very cold weather. It has been reported that cold stress reduces dry matter digestibility by 1.8% for each 10°C reduction in temperature below 20°C. The reduction in digestibility is due to increased passage rate of feed through the digestive tract. Severe cold stress affects milk production in dairy cows. If cows are not fed additional feed or the quality does not allow them to eat enough to meet their additional energy requirements, body mass will be burned to produce metabolic heat and these cows lose weight. In this condition cows have lower milk production, increased neonatal mortality and reduced growth rate in surviving calves. These cows usually have delayed return to estrus, longer day's open and poorer reproductive success (Verstegen *et al.,* 1984)

Impact of climate change on energy balance and metabolism

Thermal stress condition results in 20-30% more maintenance energy requirement ensuing reduced amount of net energy for growth and production. Increased expenditure of energy for maintenance together with reduced intake of energy results in negative energy balance, which is responsible for many of the consequences of thermal stress. Thermal stress causes reduction in blood glucose and non esterified fatty acid (NEFA) level due to reduction in hepatic glucose synthesis. Reduction of non esterified fatty acid (NEFA) level during thermal stress causes lipolysis and mobilizes adipose tissue

Impact of climate change on electrolyte and acid base balance

Increased potassium loss through skin due to increased sweating together with increased urinary sodium excretion due to lower aldosterone during thermal stress resulted in electrolyte imbalance in rumen fluid and plasma. Decreased net mineral intake due to reduced appetite and reduced absorption of minerals during hot ambient temperature results further imbalance in electrolytes and chemical reaction in blood and rumen. Similarly, hyperventilation due to increased respiratory rate reduces the level of bicarbonate (HCO3-) in blood resulting respiratory alkalosis.

Climate change and water availability for animals

Global warming is likely to intensify, accelerate the global hydrological cycle (IPCC, 2008). Changes in climate will affect water resources availability through changes in form, frequency, intensity and distribution of precipitation, soil moisture, glacier, river and groundwater flows, and lead to further deterioration of water quality. Climate change will directly affect the water demand for agriculture and livestock. Water demand for irrigation may increase as transpiration increases with higher temperatures

Innovative and practical solutions are required to manage accessible water for livestock. Most adapted species of livestock, feeding strategies (xerophytes and succulent feed resources) and water resource management could be considered in water scarcity zones. At farm level monitoring water intake for livestock is mandatory for a farm manager for maximum production. Easy access to quality and plentiful water supplies may increase livestock productivity

Effect of climate change on forage availability for livestock

Forage crops account for 60 to 90% of feedstuff input in animal production systems. Changes in the primary productivity of crops, forages and rangelands are probably the most visible effect of climate change on feed resources. Higher temperatures can cause plants to mature and complete their stages of development faster, leading to less time for accumulation of sufficient dry matter and result in yield loss for grain crops as well as forage grasses. Extreme climatic events like heat waves, droughts, floods, cyclones and heat waves have great impact on crops. Projected changes in the frequency and severity of extreme climatic events are expected to negatively impact crop yields and global food production. More frequent extreme events may lower long-term yields by directly damaging crops at specific developmental stages, such as temperature thresholds during flowering, or by making the timing of field applications more difficult, thus reducing the efficiency of farm inputs.

Use of indigenous and locally-adapted fodder grasses as well as the selection and multiplication of crop varieties and races adapted or resistant to adverse conditions will ensure adequate availability of fodder for the livestock. The selection of crops and cultivars with tolerance to stresses (e.g. high temperature, drought, flooding, high salt content) as well as resistance to insect pests and diseases helps to utilize the genetic variability in new crop varieties to cope with climate change. There are many fodder crop species already in use with a wide range of climatic adaptation and one crop can be substituted for another to suit the climate.

Nutritional intervention of different climatic stresses in domestic animals

The most effective management strategy to minimize production losses during heat stress periods is to provide a cool, comfortable environment by shading, sprinkling and/or forced air flow. Modifying the environment will result in bigger gains, or fewer losses, during heat stress periods than any dietary manipulations. Diet changes will have only a small effect on productivity and should be considered supportive and an enhancement to environmental cooling.

Dry matter intake

The primary impacts of heat stress in dairy cows are a significant reduction in voluntary feed intake. Water intake is closely related to dry matter intake (DMI) and milk yield and its intake is increased by 1.2 kg per °C increase in minimum ambient temperature, but regardless of rate of increase it is obvious that abundant water must be available at all times under hot conditions. Numerous nutritional strategies need to be implemented during this period and these include re-formulations of diets to account for reduced DMI, higher maintenance costs and metabolic heat production from various feedstuffs (West, 2003). The maintenance requirement of lactating dairy cows increases substantially as environmental temperature increases, therefore the frequency of feeding may be increased so as to increase DMI. As the intake of DM generally declines with hot weather, therefore effort should be made to increase the nutrient density of the diet. Energy density can be increased by supplementing extra grain or fat in the diet (rumen protected/rumen bypass) and a reduction in forages. Nutritionists often increase the energy or protein density of the ration during prolonged periods of heat stress. Due care should be taken if protein levels is increased during hot weather since there is an energetic cost associated with feeding of excess protein. Water intake is closely related to DMI and milk yield, indicating the influence that ambient temperature exerts on water consumption (Murphy *et al.,* 1983). Water intake increased by 1.2 kg/°C increase in minimum ambient temperature, but regardless of rate of increase it is obvious that abundant water must be available at all times under hot conditions. In addition, Milam *et al.* (1986) demonstrated that offering chilled drinking water enhanced milk yield for lactating cows by reducing body temperature through absorbed heat energy in summer stress condition.

Fiber in the diet

As dry matter intake decreases during heat stress, a concern about having adequate amounts of fiber (ADF, NDF and effective or forage NDF) in diets arises. However, fiber digestion results in a higher heat increment than digestion of fat or non-fiber carbohydrates (NFC). Acetate, the end product of fiber digestion, has a lower efficiency of utilization in the body than propionate or glucose from NFC digestion and both fiber and NFC end products are utilized less efficiently than fat (West, 2003). Thus, feeding high forage diets during summer months can add significantly to the animal's heat load. Feeding a minimum, but adequate amount of total and effective fiber should be the objective during summer months. Feed high quality forage but don't go below 18-19% ADF (acid detergent fiber). Water is the most important nutrient for the livestock.

Supplementation of fat

The advantage to including fat in the diet during hot weather is improved efficiency of energy use and greater energy intake as fats are 2.25 times greater in energy than carbohydrates. Because fat is more efficiently utilized as a source of energy than other feeds, it produces less heat during digestion and utilization. This supplemental fat can come from whole seeds such as cottonseeds or soybeans, tallow, rumen inert sources, or combinations. Most diets will contain about 3% fat (dry basis) without any high fat feeds. The next 2 to 3% fat can come from whole seeds. This results in rations with 5 to 6% fat. Anything above this should be added as rumen inert fat (commercial sources are available). Lower-fiber feedstuffs usually result in less heat increments than high-fiber feeds such as grass hays. Diets high in grain and low in fiber cause less heat stress for lactating cows because of the lower heat of digestion. However, it is critical to balance the ration properly, since milk fat may be depressed and digestive orders may result when a high grain ration is fed.

Protein Feeding

Overfeeding of protein during hot weather should be avoided because it takes energy to excrete excess nitrogen. Rations usually should be 18% protein or less on a dry basis. Only the highest producing cows will need the 18% protein ration. Similarly, do not overfeed highly degradable protein, i.e., 65 percent or greater degradable crude protein in the rumen, because this also increases the heat increment and requires more heat to be dissipated from the animal. Proper supplementation of more undegradable protein appears to be effective in reducing the heat of digestion. In poultry, requirements for protein and amino acids are independent of environmental temperature so heat stress does not affect bird performance as long as the protein requirement is met. Relatively high protein (16 to 18% CP) with adequate methionine (2% of CP) and lysine (5% of CP) level together with high energy level (11.7-12.6MJ/kg) are usually given to Leghorn hens, in hot weather situations (McNaughton *et al.,* 1977).

Mineral Supplementation

The electrolyte minerals, sodium (Na) and potassium (K), are important in the maintenance of water balance, ion balance and acid-base status of the heat stressed cow. Hot weather increases the need for certain minerals (Harris, 1992). This is due to increased sweating and urination resulting in more minerals being excreted. Potassium should be increased to at least 1.5% of dry matter, sodium to 0.45%, and magnesium to 0.35%. Magnesium may be increased already if fats are being fed. Complete minerals designed to contain the higher levels of potassium and sodium should be fed only to lactating cows since udder edema is more prevalent in dry cows receiving extra salt or potassium. Heat stressed cows responded to increasing the dietary cation anion balance (DCAB, Na + K - Cl) from 120 to 464 meq per kilogram regardless of whether Na or K was used to increase the DCAB (West *et al.,* 1992.). The calcium requirement of layers, particularly older birds, is increased at high environmental temperatures. Therefore, extra calcium should be provided at the rate of 1g/bird in the summer months in the form of oyster shell grit or limestone. Supplementation should be made over the normal dietary calcium level (3.75g/bird/d) recommended for layers. The phosphorus level in diet must not be forgotten as excessive phosphorus inhibits the release of bone calcium and the formation of calcium carbonate in shell gland, thereby reducing the shell quality. Supplementing the diet with 0.5% sodium bicarbonate or 0.3-1.0% ammonium chloride or sodium zeolites can alleviate the alkalosis caused by heat stress. The potassium requirement in bird increases from 0.4-0.6% with a rise in temperature from 25 to 38°C. A daily potassium intake of 1.8-2.3g potassium is needed by each bird for maximum weight gain under hot conditions. Additional allowances of ascorbic acid (vitamin C), vitamins A, E, and D_3 and thiamine can improve bird performance at higher temperatures.

Vitamin Supplementation

Little is known about vitamin nutrition of heat stressed cows. Some nutritionists have suggested raising levels of supplemented vitamins during heat stress. However, if you are supplementing 100,000 international units (IU) of vitamin A per day, 50,000 IU of vitamin D, and 500 IU of vitamin E, it would have better effect in animals. Cows can manufacture vitamin D with exposure to sunlight and summer is a time where we might need less supplementation. Supplementation of vitamin E and vitamin C in the diet have beneficial effect to prevent stress in animals. Supplementation of niacin during the summer increased milk production across all cows by an average of about 2 pounds per day, but cows producing over 75 pounds per day increased over 5 pounds per day (Muller *et al.,* 1986). No change in milk components was found.

Feed additives

Buffers: Feeding buffers can be beneficial during heat stress periods for two reasons. First, if fiber content of the diet is minimized and/or cows are selecting against eating forages, buffers can help prevent a low rumen pH and rumen acidosis problems. Secondly, the most common macro mineral in a buffer is usually Na, exception of K in KHCO3, which when increased in diets fed during heat stress has increased DMI and milk production.

Fungal cultures: Experiments have shown that feeding *Aspergillus oryzae* reduced heat stress in cows through lowering rectal temperatures (average 0.86° F). Milk production increased in some studies and was attributed to improved fiber digestion in the rumen (Huber *et al.,* 1994).

Conclusion

Climate change has an adverse effect on livestock produce and cause huge economic losses to farmers. Understanding of stress and the interactions of farm animals with environment will help in developing suitable strategies to overcome the adverse effect of climate change in farm animals. Nutritional management of stress, however, will be helpful in overcoming the detrimental effect, provided sound knowledge of nutritional requirements of farm animals at different age, production level and seasons are amalgamated in a holistic manner. Further research works are needed to find the nutritional requirement of stressed animals in rapidly changing climate scenario.

References

Ain Baziz, H.A., Geraert, P.A., Padilha, J.C.F., 1996. Chronic heat exposure enhances fat deposition and modifies muscle and fat partition in broiler carcasses. Poult Sci., 75:505-513.

Baumgard, L.H., Odens, L.J., Kay, J.K., 2006. Does negative energy balance (NEBAL) limit milk synthesis in early lactation? Proc. Southwest Nutr. Conf. 181-187.

Collier, R.J., Baumgard, L.H., Lock, A.L., 2005. Physiological limitations: nutrient partitioning. In: Wiseman J, Bradley R (eds) Yields of farmed Species: constraints and opportunities in the 21st century. Proceedings: 61st Easter School. Nottingham, England.

Dikshit, A.K and Birthal, P.S., 2010. India's Livestock Feed Demand: Estimates and Projections. 23(1):15-28.

Huber, J. T., Higginbotham, G., Gomez-Alarcon, R. A., Taylor, R. B., Chen, K. H., Chan, S. C., and Wu, Z. 1994. Heat stress interactions with protein, supplemental fat, and fungal cultures. J. Dairy Sci. 77:2080–2090

Intergovernmental Panel on Climate Change Secretariat (IPCC)., 2008: Technical Paper VI. In: Climate Change and Water (B.C. Bates, Z.W. Kundzewicz, S. Wu and J.P. Palutikof, eds). Geneva, IPCC Secretariat, 210 pp.

Johnson, H.D., Ragsdale, A.C., Berry, I.L., 1962. Effects of various temperature–humidity combinations on milk production of Holstein cattle. University of Missouri Agric Exp Stn Res Bull;91).

Kadzere, C.T., Murphy, M.R., Silanikove, N., 2002. Heat stress in lactating dairy cows: a review. Livest Prod Sci, 77: 59-91.

McNaughton, J.L., Kubena, L.F., Deaton, J.W., 1977. Influence of dietary protein and energy on the performance of commercial egg type pullets reared under summer conditions. Poult Sci, 56:1391-1398.

Muller, L. D., A. J. Heinrichs, J. B. Cooper, and Atkin, Y. H. 1986 Supplemental niacin for lactating cows during summer feeding. J. Dairy Sci. 69:1416.

Murphy, M. R., Davis, C. L. and McCoy G. C. 1983. Factors affecting water consumption by Holstein cows in early lactation. J. Dairy Sci. 66:35–38.

Niaber, J.A., Hahn, G.L., 2007. Livestock production system management responses to thermal challenges. International J. Biometeoro, 52:149-157.

Soren, N.M., 2013. Nutritional manipulations to optimize productivity during environmental stresses in livestock.In Environmental Stresses and Amelioration in Livestock Production. Sejian, V., Naqvi, S.M.K., Lakritz, J., Ezeji, T. and Lal, R., (Eds), Springer Publishers. pp. 181-217.

Verstegen, M.W.A., Brandsma, H.A., Mateman, G., 1984. Effect of cold thermal environment on feed requirements, growth rate and slaughter quality in pigs. Archiv für experimentelle veterinärmedizin, 38: 431-438

West, J.W., 2003. Effects of heat-stress on production in dairy cattle. J. Dairy Sci., 86:2131-2144.

West, J. W., Haydon, K. D., Mullinix, B. G. and Sandifer, T. G. 1992. Dietary cation-anion balance and cation source effects on production and acid-base status of heat-stressed cows. J. Dairy Sci., 75:2776–2786

14

Role of Nano-Selenium in Poultry Production and Reproduction

S K Mishra, K Sethy, P S Swain and S P Biswal

Profitable broiler production demands dietary nutrients including energy, protein, mineral, vitamins and additives as per the body requirement to meet high growth of the birds. Though energy and protein are the principal nutrients required for growth of birds but the essential role of minerals in poultry production cannot be ignored. Minerals have multifaceted role in the body. They are the components of bones and other soft tissues and play significant role in acid base balance, osmotic pressure regulation and membrane permeability. Trace minerals have major role in functioning of various enzymatic, metabolic and biochemical reaction for effective utilization of nutrients. They are constituents of many proteins involved in metabolism of nutrients, immune defense systems and hormone secretion pathways. Deficiency diseases, metabolic disorders, poor growth rate, low production, low hatchability and low feed efficiency are normally observed on deficiency or imbalance supply of minerals to livestock and poultry.

Along with other trace minerals, selenium is supplemented in broiler ration as an essential trace mineral required for normal growth and production of broiler birds. It was discovered in 1817 by J.J. Berzelius. Its atomic number is 34 and atomic weight is 78.96 and belongs to group VI of the periodic table. It is toxic at high dose and deficiency causes of various diseases of animals and humans. Due to role of selenium in improvement of antioxidant system, immune function, reproduction and thyroid hormone metabolism in animals, it is one of the trace mineral that is required to be added in the ration. Selenium is a component of glutathione peroxidase enzyme which serves as an antioxidant enzyme. This enzyme controls levels of hydrogen peroxide and lipid peroxides (Ahmadi *et al.*, 2018) thus protects the integrity of cell membrane phospholipids. Selenium act by controlling the body's glutathione (GSH) pool and its major selenium-containing antioxidant enzymes, glutathione peroxidase (GPx) and thioredoxin reductase (Tapiero *et al.*, 2003).

Glutathione and GPx protect the integrity of unsaturated bonds of membrane phospholipids by neutralizing the free radical attacks capable of initiating lipid oxidation of the cell (Korniluk *et al.*, 2007; Rayman, 2004). Deficiency of selenium in poultry ration has negative impact on growth of birds and feed efficiency. In selenium deficiency, pancreatic dystrophy, exudative diathesis and nutritional muscle dystrophy of the gizzard, heart and skeletal muscle are observed in poultry birds (Cantor *et al.*, 1982) thereby decreased productive and reproductive performance. Also selenium deficiency lowered immunity, fertility, embryo mortality and feathering of chickens. Increased mortality, reduced egg production, decreased hatchability, nutritional encephalomalacia, nutritional pancreaticatrophy, exudative diathesis, and muscle myopathies observed in birds due to deficiency of selenium in the diet (Tufarelli *et al.*, 2016).

As per NRC (1994), the minimum level of selenium supplementation in broilers ration is 0.15 mg/kg of diet. But to prevent loss in production of birds due to deficiency, researcher putting their effort for explore alternative selenium sources with their optimum supplemental levels. Selenium, like all biologically essential trace elements, can be toxic when provided at levels in excess of the biological requirement (Cai *et al.*, 2012). Probably, for this, European Union (2004) recommended supplementation of selenium at 0.5 mg/kg of diet in poultry for safety reason. To prevent deficiency, supplemental selenium is added in the ration to maintain productive performance and lower the chance of deficiency. Traditionally, inorganic selenium in form of sodium selenite is supplemented. But it has a narrow margin between the nutritional requirement and its toxicity (Wolffram *et al.*, 1986). In place of inorganic selenium, organic selenium is supplemented in animal feed due to higher bioavailability. The reported study of Zhan *et al.* (2007) revealed higher bioavailability and lower toxicity in pigs due to supplementation of selenomethionine and Se-enriched yeast. Further, it is supplemented in organic forms; such as selenomethionine (SeM) or Se-enriched yeast (SeY) and it was reported that organic seleneium from Se-enriched yeast had higher gut absorption and accumulation in the tissue of broilers in comparison to inorganic form (Collins *et al.*, 1993). The advantages of increased tissue concentration of selenium is to reduce drip loss from breast meat (Downs *et al.*, 2000; Naylor *et al.*, 2000) and protect the unsaturated fatty acid from damage due to peroxidation (Korniluk *et al.*, 2007). Considering the higher bioavailability of organic selenium, FDA (2000) has approved the use of it in poultry diet. To increase the bioavailability of selenium, further researches are going on to include nano-selenium in place of inorganic selenium as the bioavailability is highly correlated the physical form (Ahmadi *et al.*, 2018).

Bioavailability has been defined as: The maximal tissue mineral utilization for biological functions in metabolism, based on the amount of ingested mineral (Kirchgessner *et al.*, 1993). Bioavailability, in reference to trace minerals, is defined as the proportion of the ingested element that is absorbed, transported to the site of action, and converted to a physiologically active form (O 'Dell, 1983). The amount of mineral required for supplementation is dependent upon the bioavailability of a mineral source and higher bioavailability could reduce the amount of a mineral that is added to a diet to meet the dietary requirements, which would reduce the amount of mineral excreted (Cheng *et al.*, 1998). The physical form of selenium is the key factor for deciding the bioavailability of selenium. Nano mineral (in this context, nano selenium) has higher bioavailability due to large surface, excellent surface activity, good catalytic efficiency, high absorbing ability, and low toxicity (Wang *et al.*, 2007; Zhang *et al.*, 2008).

Nanotechnology can be defined as manipulation of (designing, synthesis) particle with dimensions less than a micron to that of individual atom. Nanotechnology is undoubtedly one of the mostimportant technologies of the 21st century as it has the potential to provide the tools and the research to change the future of food and feed technology. It has been referred to as the next industrial revolution (Schmidt, 2009). The particle size of nano minerals are in the range of 1-100 nano meter (Feng *et al.*, 2009). Nanoparticles present a higher surface volume with decreasing size of the particle. The physical, chemical, electrical, optical, mechanical, and magnetic properties of the nanoparticles are quite different from those present at a greater scale (Buzea *et al.*, 2007). Due to nano meter size of selenium in the redox state of zero, higher stability and solubility of selenium, it has drawn attention of many research workers There are several methods for preparation of nano particles but it is important to determine the size, shape, surface area, phase constitution, and micro-structural features. The size and shape of powder particles may be determined accurately with use of either scanning electron microscopy (SEM) for relatively coarse powders or transmission electron microscopy (TEM) for fine powders.

Effect of nano selenium supplementation on the performance, feed efficiency and feed consumption of broilers was reported by many research workers. Contradictory reports were available as regard to the growth performance of broilers on nano selenium supplementation. Ahmadi *et al.* (2018) supplemented nano selenium at 0.1 mg/kg of feed, 0.2 mg/kg of feed, 0.3 mg/kg of feed, 0.4 mg/kg of feed and 0.5 mg/kg of feed in broiler ration and compared the body weight with control. They reported that supplementation of nano-selenium significantly improved weight gains and feed conversion ratio in during 1st -

21st day and 22nd - 42nd day periods. Similar findings of higher body weight in nano selenium supplemented groups were also reported by Zhou and Wang (2011) in Guangxi Yellow chickens. They supplemented nano Se (0, 0.1, 0.3 and 0.5 mg/kg diet) in chicken diet and reported higher body weight gain and recommended that 0.3 mg/kg nano-Se in chicken diet was optimum for better body weight gain performance of birds. Supplementation of nano selenium at higher levels affected the growth of broilers. Higher selenium levels in the rations can negatively affect animal health (Suchy *et al.*, 2014). Cai *et al.* (2012) reported no significant improvement in body weight on increasing nano selenium levels from 0.3 to 2.0 mg nano-selenium per kg of diet. Zhou and Wang (2011) reported significantly lower body weight of broilers in 0.1 ppm nano supplemented group than that of 0.3 ppm and 0.5 ppm nano supplemented groups. They also observed no significant difference in body weight of broiler in 0.3 and 0.5 nano supplemented groups. The reduced weight gain in broilers might be due to toxic effect of the selenium at higher level of supplementation. Prasoon *et al.* (2018) reported that supplementation of sodium selenite at 0.3 ppm and nano selenium at 0.005, 0.15 and 0.35 ppm levels in chicken diet had no significant effect on the feed consumption and FCR at 21st, 42nd and 56th day in comparison to control. Similar results were also reported by Ahmadi *et al.* (2018). Cai *et al.* (2012) reported no significant differences in feed intake broilers fed with diets supplemented nano selenium at 0.3 to 2.0 ppm levels. Ahmadi *et al.* (2018) reported significantly higher FCR in control group than that of nano selenium supplemented group. Cai *etal.* (2012) reported no significant differences in feed conversion ratio of broilers fed with diets supplemented nano selenium at 0.3 to 2.0 ppm levels. Poultry birds are subjected to many kind of stress, which may hinder their development, growth, production and immunity. During stress condition, muscle proteins are catabolised by proteasomes and selenium is released, which would be a source of slenium for enzymes like glutathione peroxidase and thioredoxin reductase. These enzymes prevent free radicals produced as intermediate products during the course of metabolism process, which are responsible for decreased growth and reproductive performances. Thus, Se supplementation could alleviate the negative effect of stress on egg production in laying hens, which might be responsible for increased egg production by combating various stress factors. Selenium increased the laying performance of hens as per the reported works of El-Latif (1999), Ganpule and Manjunatha (2003) and Sahin *et al.* (2003). Study of Scheideler *et al.* (2010) revealed that level of dietary selenium significantly affected the rate of egg production. But Pavlovi *et al.* (2009) did not observe any differences in egg production during the first 8 weeks of dietary selenium administration in laying hens, but in the last 8 weeks selenium yeast supplementation, they observed increased egg production compared to control and sodium selenite. Mohapatra *et al.* (2014) reported that egg production of layer birds was affected by dietary nano-Se level. Effect on serum biochemical indices and hematological profile of broilers on nano selenium supplementation of different reported works are elucidated below. Ahmadi *et al.* (2018) reported no significant effect on plasma glucose and total protein concentrations on supplementation of nano-selenium in broilers. Okunlola *et al.* (2015) in their experiment fed selenium at 0.3 and 0.5 mg/kg levels and reported no significant differences in serum total protein, albumin, urea and creatine levels in birds. Rostamabad *et al.* (2017) reported that supplementation of nano-Se linearly decreased the plasma concentrations of AST before heat exposure. Okunlola *et al.* (2015) in their experiment fed selenium at 0.3 and 0.5 mg/kg levels and reported no significant differences ($p>0.05$) in packed cell volume, heamoglobin, red blood cell and white blood cell among different treated groups.

The principal role of selenium supplementation is for improvement in antioxidant status and immunity of birds. Selenium is essential for maintaining the immunity of the body and its deficiency lower both cellular and humoral immunity (Arthur *et al.*, 2003). Also through proliferation of activated T lymphocytes, it stimulates the immune system (Rayman, 2000). Different workers reported their findings as regard to the effect of nano selenium supplementation in birds. Zhang *et al.* (2012) reported that the selenium supplementation in the chicken diets improved the immunological parameters. Bagheri *et al.* (2015) reported that chickens supplemented with nano-selenium had significantly higher immunity response than that of control group. Ahmadi *et al.* (2018) reported that supplementation of nano selenium significantly increases compared to the control for anti-Newcastle disease hemagglutination-inhibitiontiter after the dietary supplementation with the nano-selenium. Cai *et al* (2012) reported significant increased in IgM level in groups supplemented with 0.3 to 1.0 mg/kg of nano-selenium and birds fed 0.30 mg/kg of nano-Se exhibited the highest IgG and IgM levels.

Glutathione peroxidase is an enzyme transforming the toxic and carcinogenic hydrogen peroxide to harmless water and oxygen and its activation requires small amounts of selenium (selenocysteine), probably substituting sulphur in the glutathione molecule and causing development of modified enzyme GPx4 (Suchy *et al.*, 2014). Glutathone perioxidase protect the integrity of unsaturated bonds of membrane phospholipids by extinguishing free radical attacks capable lipid oxidation (Korniluk *et al.*, 2007; Rayman, 2004). Zhou and Wang (2011) reported that chickens supplemented with Nano-Se recorded significantly higher serum GSH-Px activities than that of control. Cai *et al.* (2012) reported that birds supplemented with 0.3, 0.5, 1.0, and 2.0 mg/

kg of nano-Se showed increased glutathione peroxidase activity compared with the control group. Cai *et al.* (2012) reported that the total superoxide dismutase activity, free radical inhibition and the content of glutathione and malondialdehyde were not affected by dietary nano-selenium levels. Aparna and Karunakaran (2016) reported that that birds supplemented with 0.1875 mg/kg selenium nanoparticles had significantly higher superoxide dismutase activity and the superoxide dismutase activity reduced when they increased the level to 0.225 mg/kg.

As regard to the selenium deposition in the tissues, different workers reported their finding and are summarized in this chapter. Hu *et al.* (2012) also reported that supplementation nano-selenium had higher selenium retention in liver and muscle as compared to selenite and opined that might be partly the reason for the lower toxicity of nano-selenium compared with selenite. The different retention of nano-selenium and sodium selenite was probably related to the different absorption process and metabolic pathways (Hu *et al.*, 2012). This might be due to higher bioavailability of nano selenium. It has been reported that nano particle show new characteristics of transport and uptake and exhibit higher absorption efficiencies (Davda and Labhasetwar, 2002; Chithrani and Chan, 2007; Zha *et al.*, 2008; Liao *et al.*, 2010). Certain relationship must be existing between nano-Se metabolism and liver function and meat quality for which the liver and muscle selenium content increased on supplementation of nano selenium (Cai *et al.*, 2012) in broilers.

References

Ahmadi, M., Ahmadian, A. and Seidavi, A.R. 2018. Effect of Different levels of nano-selenium on performance, blood parameters, immunity and carcass characteristics of broiler chickens. Poultry Science Journal, 6 (1): 99-108.

Aparna, N. and Karunakaran, R. 2016. Effect of Selenium nanoparticles supplementation on oxidation resistance of broiler, chicken, Indian Journal of Science and Technology, Online. 9 (S1), DOI: 10.17485/ijst/2016/v9iS1/106334.

Arthur, J.R., McKenzie, R.C. and Beckett, G.J. 2003. Selenium in the immune system, Journal of Nutrition, 133: 1457–1459.

Bagheri, M., Golchin-gelehdooni, S., Mohamadi, M. and Tabidian, A. 2015. Comparative effects of nano, mineral and organic selenium on growth performance, immunity responses and total antioxidant activity in broiler chickens, Inernational Journal of Biology, Phramacy and Allied Science, 4 (2): 583-595.

Buzea, C., Pacheco, I.I. and Robbie, K. 2007. Nanomaterials and nanoparticles: Sources and toxicity. Biointerphases, 2(4):17-71

Cai, S.J., Wu, C.X., Gong, L.M., Song, T., Wu, H. and Zhang, L.Y. 2012. Effects of nano-selenium on performance, meat quality, immune function, oxidation resistance, and tissue selenium content in broilers, Poultry Science, 91: 2532–2539.

Cantor, A.H., Moorehead, P.D. and Musser, M.A. 1982. Comparative effects of sodium selenite and selenomethionine upon nutritional muscular dystrophy, selenium- dependent glutathione peroxidase, and tissue selenium concentrations of turkey poults, Poultry Science, 61: 478–484.

Cheng, J., Kornegay, E. T. and Schell, T. 1998. Influence of dietary lysine on the utilization of zinc from zinc sulphate and a zinc-lysine complex by young pigs, Alliance Of Crop, Soil and Environmental Science Societies Digital Library, 76 (4): 1064-1074.

Chithrani, B.D. and Chan, W.C.W. 2007. Elucidating the mechanism of cellular uptake and removal of protein-coated gold nanoparticles of different sizes and shapes. Nano Lett. 7: 1542–1550.

Davda J and Labhasetwar V. 2000. Characterization of nanoparticle uptake by endothelial cells, Intentional Journal of Pharmacology, 233: 51–59.

Downs, K.M., J.B. Hess and S.F. Bilgili, 2000. Selenium source effect on broiler carcass characteristics, meat quality and drip loss, Journal of Applied Animal Research, 18: 61-71.

El-Latif, S.A.A. 1999. Nutritional interrelationships of vitamin E and selenium on laying Japanese quail, Egyptian Journal of Nutrition Feeds, 2: 711-718. Cited in Poultry Abstract, 2000, 26(9): 2416. FDA. 2000.

FDA approves food additive petition for selenium yeast. FDA Veterinarian Newsletter, U. S. Food and Drug Administration, Washington, D.C. pp. 10.

Feng, M., Wang, Z.S., Zhou, A.G. and Ai. D.W. 2009. The effects of different sizes of nanometer zinc oxide on the proliferation and cell integrity of mice duodenum- epithelial cells in primary culture. Pakistan Journal of Nutrition. 8(8):1164-116

Ganpule, S.P. and Manjunatha, B.P. 2003. Antioxidant nutrition starts with breeders, Feed mix, 11: 14-16.

Hu, C.H., Li, Y.L., Xiong, L., Zhang, H.M., Song, J. and Xia, M.S. 2012. Comparative effects of nano elemental selenium and sodium selenite on selenium retention in broiler chickens, Animal Feed Science and Technology, 177 (3-4): 204-210

Kirchgessner, M., Windisch, W. and Weigand, E. 1993.True bioavailability of zinc and manganese by isotope dilution technique, Bioavailability, 93: 213-222.

Korniluk, K., Czauderna, M. and Kowalczyk, J. 2007. The influence of dietary conjugated linoleic acid isomers and selenized yeast on the fatty acid profile of the spleen, pancreas and kidneys of rats, Journal of Animal and Feed Sciences, 16: 121-139.

Liao CD, Hung WL, Jan KC, Yeh AI, Ho CT and Hwang LS. 2010. Nano/sub microsized gnan glycosides from sesame meal exhibit higher transport and absorption efficiency in Caco-2 cell monolayer, Food Chemistry, 119: 896–902.

Mohapatra, P., Swain, R.K., Mishra, S.K., Behera, T., Swain, P., Mishra, S.S., Behura, N.C., Sabat, S.C., Sethy, K., Dhama, K.and Jayasankar, P. 2014. Effects of Dietary Nano-Selenium on Tissue Selenium Deposition, Antioxidant Status and Immune Functions in Layer chicks, International Journal Pharmacology, 10: 160-167.

Naylor, A. J., Choct, M. and Jacques. K.A. 2000. Effects of selenium source and level on performance and meat quality in male broilers, Poultry Science, 79 (Suppl. 1):117.

NRC. 1994. National Research Council, Nutrient Requirements of Poultry: 9th Revised Edition, National Academy Press, Washington, DC.

O'Dell, B. L. 1983. Bioavailability of essential and toxic trace elements. Fed. Proc. 42:1714. Okunlola, D.O., Akande, T.O. and Nuga, H.A. 2015. Haematological and serum characteristics of broiler birds fed diets supplemented with varying levels of selenium powder, Journal of Biology Agriculture and Healthcare, 5 (1): 107-110.

Pavlovi, Z., Mileti, I., Joki, Z., Pavlovski, Z., Skrbi, Z. and Sobaji, S. 2010. The effect of level and source of dietary selenium supplementation on eggshell quality, Biological Trace Elements Research, 133: 197-202.

Prasoon, S., Jayanaik, , Malathi, V., Nagaraja, C.S. and. Narayanaswami, H.D. 2018. Effects of Dietary Supplementation of Inorganic, Organic and Nano Selenium on Antioxidant Status of Giriraja Chicken. Online. https://doi.org/10.20546/ijcmas.708.242.

Rayman M.P. 2000. The importance of selenium to human health, Lancet, 356: 233–241. Rayman, M.P. 2004. The use of high-selenium yeast to raise selenium status: how does it measure up?, British Journal of Nutrition, 92 : 557–574.

Rostamabad, Morteza afdari-, Seyyed, Javad Hosseini-Vashan, Ali Hossein, Perai and Hadi, Sarir. 2017. Nanoselenium Supplementation of Heat-Stressed Broilers: Effects on Performance, Carcass Characteristics, Blood Metabolites, Immune Response, Antioxidant Status, and Jejunal Morphology, Biological Trace Elements Research, 178 : 105–116.

Sahin, N., Sahin, K. and Onderci, M. 2003. Vitamin E and selenium supplementation to alleviate cold-stress associated deterioration in egg quality and egg yolk mineral concentrations of Japanese quails, Biological Trace Elements Research, 96: 179-189. Scheideler, S.E., Weber, P. and Monsalve, D. 2010. Supplemental vitamin E and selenium effects on egg production, egg quality, and egg deposition of α-tocopherol and selenium, Journal of Applied Poultry Research, 19: 354–360.

Schmidt, Charles W. 2009. Nanotechnology-Related Environment, Health, and Safety Research: Examining the National Strategy. Environmental Health Perspectives, 117(4): A158–A161.

Suchy, P., Strakova, E. and Herzig, I. 2014. Selenium in poultry nutrition: a review, Czech Journal of Animal Science, 59 (11): 495–503

Tapiero, H, Townsend, D.M. and Tew, K.D. 2003. The antioxidant role of selenium and selenocompounds, Biomed. Pharmacotherapy, 57: 134-144

Tufarelli, V., Cazzato, E., Ceci, E. and Laudadio, V. 2016. Selenium-fertilized tritordeum (× Tritordeum Ascherson et Graebner) as dietary selenium supplement in laying hens: effects on egg quality, Biological Trace Element Research, 173(1): 219-24.

Wang, H.L., Zhang, J.S. and Yu, H.Q. 2007. Elemental selenium at nano size possesses lower toxicity without compromising the fundamental effect on selenoenzymes: Comparison with selenomethionine in mice, Free Radical Biology and Medicine, 42 (10): 1524-1533.

Wolffram, S., Anliker, E. and Scharrer, E. 1986. Uptake of selenate and selenite by isolated intestinal brush border membrane vesicles from pig, sheep and rat. Biol. Trace Elem. Res. 10, 293-306.

Zha, L.Y., Xu, Z.R., Wang, M.Q. and Gu, L.Y. 2008. Chromium nanoparticle exhibits higherabsorption efficiency than chromium picolinate and chromium chloride in Caco-2 cell monolayers, Journal of Animal Physiology and Animal Nutrition, 92: 131–140.

Zhan, X.A., Wang, M., Zhao, R.Q., Li, W.F. and Xu, Z.R. 2007. Effects of different selenium source on selenium distribution, loin quality and antioxidant status in finishing pigs, Animal Feed Science and Technology, 132: 202–211.

Zhang, Z.W., Wang, Q.H., Zhang, J.L., Li, S., Wang, X.L. and Xu, S.W. 2012. Effects of oxidative stress on immunosuppression induced by selenium deficiency in chickens, Biological Trace Element Research, 149: 352–361.

Zhang, J.S., Wang, X.F. and Xu, T.W. 2008. Elemental selenium at nano size (Nano-Se) as a potential chemopreventive agent with reduced risk of selenium toxicity: comparison with Se-methylselenocysteine in mice, Toxicological Sciences, 101: 22– 31.

Zhou, X. and Wang, Y. 2011. Influence of dietary nano elemental selenium on, growth performance, tissue selenium distribution, meat quality, and glutathione peroxidase activity in Guangxi Yellow chicken, Poultry Science, 90 (3): 680–686.

15

Climate Change: Mitigation and Adaptation Strategies in Agriculture and Livestock Sector

H N Malik, A Panda, D P Das, B C Das, K K Sardar

Introduction

Climate change is one of the important areas of concern for world to ensure food and nutrition security to the growing population. The impacts of climate change are global but development countries like India are highly vulnerable as large population depends upon agriculture and allied service. As per the latest report, the global average temperature rises is 0.99°C (NASA, 2016) since pre-industrial time (1850). The year 2016 ranks as the warmest year, 16 of the 17 warmest years in the 136 years record all have occurred since 2001. The predicted temperature rise for India is in the range 0.5-1.2°C by 2020, 0.88-3.16°C by 2050 and 1.56-5.44°C by the year 2080. Studies in India showed significant negative impacts of climate change, predicted to reduce yields by 4.5 to 9.0%, depending on the magnitude and distribution of warming. Agriculture sector is contributing about 17.4% of India's GDP, a 4.5-9.0% negative impact on production implies a cost of climate change to be roughly up to 1.5% GDP per year. Therefore, Govt. of India has accorded high priority on research and development to cope with the climate change in general and agriculture and allied sector. The Prime Ministers National Action Plan on Climate Change has Identified Agriculture and allied sector as one of the 8 national Missions. Agriculture and the future of global food security figure very importantly in climate change negotiations. As stated in Article II of the United Nations Framework Convention on Climate Change (UNFCCC), the goal is to ensure stabilization of greenhouse gas concentrations in the atmosphere at a level that would prevent "dangerous anthropogenic interference with the climate system". With this background to meet the challenges of sustaining domestic food production in the face of changing climate and to generate information on adaptation and mitigation in agriculture to contribute to global for a like UNFCCC, the Indian Council of Agricultural Research (ICAR) launched a flagship network project National Initiative on Climate Resilient Agriculture (NICRA) during XI plan in February 2011, and during XII Plan it is referred as National Innovations in Climate Resilient Agriculture (NICRA).

Key Impacts of Climate Change

Global warming

Early decades of the twenty-first century will see a moderate warming of 1-2C, resulting in reduced crop yields in seasonally dry and tropical regions, while crop and pasture yields in temperate regions may benefit. Further warming in the second half of the century will negatively affect all regions, although agriculture in many developing countries in semi-tropical and tropical regions will bear the brunt of the effects.

Extreme climate events

Increased frequency and severity of extreme climate events, such as more heat stress, droughts and flooding, is expected in coming decades due to climate change. It will increase negative impacts on agriculture, forestry and fisheries in all regions. In particular, it will modify the risks of fires, and pest and pathogen outbreaks, with negative consequences for food, fibre and forestry.

Undernourishment

The number of undernourished is likely to increase by 5-170 million people by 2080, with respect to a baseline with no climate change. Even small amounts of warming will increase risk of hunger in poor developing countries, due to negative impacts on food production and availability. Most of the increases are projected in sub-Saharan Africa (Easterling *et al.,* 2007).

Food stability, utilization and access

Additional negative impacts of climate change on food security, with the potential of reducing access to and utilization of food in many regions already vulnerable today, are expected but have not been quantified. In particular stability of food supply is likely to be disrupted by more frequent and severe climate extremes. Utilization of food may be affected negatively by increases in crop, livestock and human pests and diseases, as well as by reduced water availability and water quality, of importance for food preparation.

Mitigation and Adaptation

When it comes to tackling climate change to prevent the impacts it causes in the different systems of the planet, the human being applies two types of measures: mitigation and adaptation. Mitigation measures are those actions that are taken to reduce and curb greenhouse gas emissions, while adaptation measures are based on reducing vulnerability to the effects of climate change. Mitigation therefore, attends to the causes of climate change, while adaptation addresses its impacts. How to mitigate climate change? These are some of the mitigation measures that can be taken to avoid the increase of pollutant emissions: Practice Energy efficiency Greater use of renewable energy Electrification of industrial processes Efficient means of transport implementation: electric public transport bicycle, shared cars ... Carbon tax and emissions markets. Adaptation to climate change: In terms of adaptation measures, there are several actions that help reducing vulnerability to the consequences of climate change: More secure facility locations and infrastructures Landscape restoration (natural landscape) and reforestation Flexible and diverse cultivation to be prepared for natural catastrophes Research and development on possible catastrophes, temperature behavior, etc.

Adaptation and Mitigation Strategies in Agriculture and Livestock Sector

Natural Resource Management

- Practice in-situ moisture conservation RCT method (BBF/Ridge&furrow/ contour trenching/mulching/conservation furrow/bunding etc.).
- Facilitate water harvesting and recycling for supplemental irrigation (Community ponds/farmponds/ jalkunds/ checkdams/polybag check dams/wells etc.)
- Conservation tillage where appropriate like zero tillage/ minimum tillage etc.
- Improved drainage in flood prone areas
- Artificial ground water recharge
- Water saving irrigation methods (Drip/sprinkler/rain gun etc.)
- Crop residue incorporation instead of burning

Crop Production

- Altering inputs, varieties and species for increased resistance to heat shock and drought, flooding and salinization.
- Advancement of planting dates of *rabi* crops in areas with terminal heat stress
- Water saving paddy cultivation methods (SRI, aerobic, direct seeding)
- Frost management in horticultural crops through fumigation

- Community nurseries for delayed monsoon
- Custom hiring centres for timely planting
- Location specific intercropping systems with high sustainable yield index
- Crop diversification
- Establishment of seed bank in village level
- Expand climate literacy through a village level weather station
- Altering fertilizer rates to maintain grain or fruit quality; altering amounts and timing of irrigation and other water management;
- Altering the timing or location of cropping activities.
- Managing river basins for more efficient delivery of irrigation services and prevent water logging, erosion and nutrient leaching; making wider use of technologies to “harvest” water and conserve soil moisture; use and transport water more effectively.
- Diversifying income through the integration of activities such as livestock raising, fish production in rice paddies, etc.
- Making wider use of integrated pest and pathogen management, developing and using varieties and species resistant to pests and diseases; improving quarantine capabilities and monitoring programmes.
- Increasing use of climate forecasting to reduce production risk (Howden *et al.*, 2007).
- Undertaking changes in forest management, including hardwood/softwood species mix, timber growth and harvesting patterns, rotation periods; shifting to species or areas more productive under new climatic conditions, planning landscapes to minimize fire and insect damage, adjusting fire management systems; initiating prescribed burning that reduces forest vulnerability to increased insect outbreaks as a non-chemical insect control; and adjusting harvesting schedules.
- Introducing forest conservation, agroforestry and forest-based enterprises for diversification of rural incomes.

Livestock Sector

- Use of community lands for fodder production during droughts / floods
- Introduction of new fodder crops or new varieties
- Improved fodder/feed storage methods like silage and hays
- Improved shelters for reducing heat stress/ cold stress/ water logging/ flood and diseases in livestock

- Introduction of improved breeds (drought tolerant)
- Improved feeding like location specific mineral mixtures or mineral bricks
- Feeding of straw with urea-molasses and mineral block
- Supplementation of livestock feed with Vitamin-E and antioxidant
- Practice thorn less cactus cultivation as fodder for livestock
- Ensure proper vaccination to livestock
- Establishment of fodder bank in village level
- Matching livestock stocking rates with pasture production, altered pasture rotation, modification of grazing times, alteration of forage and animal species/breeds, integration within livestock/crop systems including the use of adapted forage crops, re-assessing fertilizer applications and the use of supplementary feeds and concentrates.
- Altering catch size and effort and improving the environment where breeding occurs; reducing the level of fishing in order to sustain yields of fish stocks.

Long term strategies (Current to 2050 and beyond)

Climate change response strategies that include a focus on rural livelihoods and food security need to be framed within comprehensive and socio-economic development policies that transcend current models. This requires strategies that aim at conserving natural resources and limit carbon footprints in the face of large demand for food, water and other environmental services. In particular, it should be recognized that beyond 2050, land-based mitigation from avoided deforestation, agroforestry and soil carbon sequestration in agricultural soils, necessary to stabilize emissions in the short term, would have largely reached their potential. New green technologies and land management options will then be necessary to mitigate emissions of greenhouse gases while making agriculture carbon neutral. It is critical to reducing the overall carbon footprint of agriculture and forestry, and maintaining soil and water resources in the face of socio-economic pressures and climate change while, at the same time, safeguarding food security through enhanced rural livelihoods. This will require attention to complex spatial, temporal and sector interactions that characterize the entire food and ecosystem service chain starting from local production and considering safety and health controls and transport, infrastructure, processing and consumption patterns. Open economic trade, reduced subsidies and income diversification, coupled with potential new income for rural populations from enhanced environmental services (including carbon and energy), likely represent important steps for sustaining successful adaptation and mitigation strategies.

References

Easterling, W.E., Aggarwal, P.K., Batima, P.,Brander, K.M., Erda, L., Howden, S.M., Kirilenko, A., Morton, J., Soussana, J.-F., Schmidhuber, J. &Tubiello, F.N. 2007. Food, fibre and forest products. Climate Change 2007: Impacts, Adaptation and Vulnerability. Contribution of WG II to the Fourth Assessment Report of the IPCC, M.L. Parry, O.F. Canziani, J.P. Palutikof, P.J. van der Linden and C.E. Hanson, eds., Cambridge University Press, Cambridge, UK, 273-313.

Howden, M., Soussana, J.F. &Tubiello, F.N. 2007. Adaptation strategies for climate change. Proc. Nat. Ac. Sciences 104:19691-19698.

IPCC. 2007. Mitigation of Climate Change. Assessment Report 4, Working Group III, InterGovernmental Panel on Climate Change.

UNFCCC. 2007. Investment And Financial Flows To Address Climate Change. UNFCCC, Bonn, Germany

16

Gynaecological Interventions to Mitigate the Adverse Climatic Factor in High Yielding Dairy Animals

P C Mishra

Introduction

Climate change and pollution are the two major concerns that retard growth in animal sector particularly in a developing country like India. Pollution is directly contributed by human beings, while the former is also an after effect of continuous abuse of nature and natural resources. Hence these two areas are to be looked into seriously for country's growth in general and animal production in particular.

Stress Effect

Stress is nothing but inability of an animal to cope with its environment, a phenomenon that is often ends up with a failure to achieve the full genetic potential. Stress may be from any origin and can even be a disease itself. There exists a variety of endocrine regulatory points of stress which limits the efficiency of production and reproduction. A simple transport produces an immediate constant increase in arginine, vasopressin and corticotrophin-releasing hormone secretion in animals. However, adenocorticotrophic hormone reaches a maximum in the first hour, while cortisol is highest during the second hour. Contrastingly, the hypothalamo–pituitary–adrenal response is delayed occurring only after glucose decreases below a threshold in an insulin treated animal. Negative feedback effects appear to operate mainly at the pituitary level during transport but at the hypothalamus during hypoglycaemia.

Climate effect on reproduction

Role of climate change on reproduction is of immense importance in high yielding cattle. The stress caused due to lactation, nutrition and management is added with that of the climate particulary when the onset is sudden. Among the most common climatic changes are thermal (either low or high), humidity light intensity, day light length, flood and cyclone. Starting from calf-hood it can adversely affect puberty, sexual maturity, estrus, follicular dynamics ovulation, fertilization, implantation, conception, growth and survivability of embryo/, parturition, lactation, involution and finally resumption of cycle. All these events are directly or indirectly controlled by HPA axis where a cascade of hormones is involved. For optimum endocrine secretions from different sources optimum time (age), nutrition and favourable climatic conditions are necessary. A slight deviation in any of the three may end with aberrations. In normal physiological balance or homeostasis, we expect every secretion to be normal. However any stress on the body might lead to impaired secretion and function resulting in low peak and low pulse. The major reoductive hormones like GnRH, FSH, LH, E_2, P_4, PG, Oxytocin, Inhibin, Relaxin and Melatonin are heavily interrelated. There is positive or negative feed back mechanism existing among them. Synergistic or antagonistic effect, can also upregulate or down regulate of one or more hormones. Besides, there are number of related hormones like GH, TSH, T_3, T_4 and IGF which are more general in action, but can influence reproduction. Since production of milk is an after effect of reproduction, it is directly linked to farmer economy. Moreover, high producing animals are of more concern because they contribute most to production.

There are endocrine evidences to show that stressors interfere with precise timings of reproductive hormone release within the follicular phase. Transport, or insulin, reduce the frequency and amplitude of gonadotrophin-releasing hormone and LH pulses, suggesting that these stressors exert effects at the hypothalamus or higher centres in the brain. Both stressors also delay the onset of the luteinising hormone (LH) surge. Preliminary results suggest that opioids mediate these effects but progesterone/glucocorticoid receptors are not involved. There is also evidence to support effects at pituitary level because exogenous ACTH, or transport, reduce the amount of LH released by challenges with GnRH. The reduction in endogenous GnRH/LH secretion ultimately deprives the ovarian follicle of adequate gonadotrophin support leading to reduced oestradiol production by slower growing follicles. Thus, there is a level of interference by stressors at the ovary. Reproduction is such an important physiological system that animals have to ensure that they can respond to their surroundings but yet have normal or reduced fertility.

Role of endocrine system

When stress is due to a changed environment, the same cascade of events may follow. Reproductive fitness may be regarded as the most important criteria for studying or evaluating animal adaptation. Body systems activated by stress are considered to influence reproduction by altering the activities of the HPA axis. Activation of stress pathways may directly affect the activity of Gonadotropin-releasing hormone neurons within the hypothalamus or higher neural centers which in turn affects the synthesis or secretion of GnRH into the hypophysial portal blood. It is also possible that stress directly influences the responsiveness of gonadotrophin cells in the anterior pituitary gland through GnRH action. A still potential action of stress is to alter the feedback actions of sex steroids in the hypothalamus or pituitary and inhibin in the anterior pituitary gland. Reproduction processes in animals may be impacted during heat exposure and glucocorticoids are paramount in mediating the inhibitory effects of stress on reproduction. Glucocorticoids are capable of enhancing the negative feedback effects of estradiol and reducing the stimulation of GnRH receptor expression by estrogen. Glucocorticoids may also exert direct inhibitory effects on gonadal steroid secretion and sensitivity of target tissues to sex steroids. Heat stress influences estrous length and characteristics and embryo survivability and production. The birth weights of newborns of heat stressed animals are generally lower than the unstressed animals. Impairment of placental size and function may contribute for it. Secretion of the hormones regulating reproductive tract function may also be altered by heat stress. Further it can inhibit 3-beta-hydroxysteroid dehydrogenase thereby minimizing progesterone secretion from luteal cells. Aromatase is an enzyme that converts androgens into estrogens and is present in the granulosa cells. Inhibition of the expression of this enzyme, may induce follicular atresia and consequently anestrum. Effects of steroid hormones on reproductive tract tissue could be reduced during exposure to heat stress due to increased synthesis of certain proteins. Further, increased amount of these stresses increases secretion of prostaglandin and reduce the secretion of interferon tau which affects the maternal recognition of pregnancy. In male, heat stress adversely affects spermatogenesis by inhibiting the proliferation of spermatocytes.

Mammals do have an extraordinary biochemical and endocrinological diversity to ensure reproduction. But synchrony between maternal and embryonic development is obligatory to establish pregnancy, which is true for all species. Analogous to embryonic development uterine secretions change quantitatively and qualitatively during early pregnancy. Among uterine secretory components, the proteins have been investigated most extensively and in many species. Most are of blood serum origin. But uterine proteins synthesized locally by the endometrium, also contribute to uterine luminal fluid during the preimplantational period. Typical protein pattern changes have been observeds as well as variations in enzyme activities. In some species pregnancy specific proteins have been identified representing embryonic-maternal interaction in early pregnancy. Uterine secretory activity is controlled by maternal hormones. An inadequate uterine milieu provokes embryonic mortality which is of importance in humans as well as in farm animals. The uterine environment serves various functions in mammals. It allows fertilization by enabling spermatozoa to ascend to the site of fertilization in the oviduct; provides adequate nutrients for the different stages of embryonic development from arrival in the uterus after tubal transport to implantation; maintains an appropriate milieu for the physico-chemical integrity of embryonic structure and fulfills immunological requirements (both immunosuppressive and antibacterial requirements. Under stressful conditions these mechanisms fail to occur normally.

Due to poor body immunity under stress, pathogenic microorganisms flare up and cause different diseases. The increased populations of the normal microflora or opportunistic ones also can be pathogenic and cause disease.

Amelioration

Since the gradual or sudden change n climatic factors are inevitable, they are to be managed to have a least effect. The following are a few which can be resorted to.

1. The overall comfort of the animal to be looked into to minimize the stress.
2. Since there is fall in feed intake, the quality of the feed should be increased. By pass fat and by pass protein with vitamins and antioxidants like vitamine A & E are a better preference to prevent much deviation in the micro environment of reproductive tract.
3. There should be regular provision of quality mineral supplementation. The nano-mineral or chelated ones are better.
4. Provision of cooling devices like fan, cooler, blower, air conditioner and sprinkler are better alternatives besides better aeration, ventilation with more floor space.
5. Use of prostaglandin at the time of AI increases conception rate in heat stressed animals.
6. Timely insemination protocols to assure ovulation and prevent suppressed estrus should be preferred.
7. Stabilization of corpus luteum by additional doses of GnRH or hCG can be practiced.
8. Estradiol based protocols can provide pronouncement of estrus while LH/ HCG based protocols can assure timely ovulation thereby increases conception rate.

9. Progesterone implants should always be preferred in post partum acyclicity for inducing ovulatory estrus
10. In normal inseminations, close watch for estrus signs and AI in late evening particularly in buffaloes works better.
11. Provision of round the clock good quality water
12. Use of immunomodulators and anthelmintics to improve immunity.
13. The sanitation and hygiene should be improved to reduce rate of infection or microbial invasion.

17

Innovative Adaptation Strategies for Control of Green House Gas(GHG) Production Through Smart-Animal Agriculture

Basudev Behera

The extreme weather events being experienced at a greater frequency now a days are due to accelerated climate change caused mostly due to global warming. Global warming is basically a change in the climatic conditions of the Earth, brought about by a considerable rise in the near-surface temperature of the planet. Earth's surface and the troposphere become warmer due to absorption of infrared radiation emitted by the Earth's surface by Green House Gases(GHGs). The important green house gases include carbon dioxide(CO_2),methane(CH_4) nitrous oxide(N_2O), chloro-fluoro-carbons(CFCs), ozone(O_3) and water vapour(H_2O). Various climate models predict major changes in our weather patterns, environment, and way of life in absence of efforts for substantial reductions in GHG emissions. The CO_2 gas concentration of atmosphere has increased to 385 ppm as against the normal concentration of 300 ppm. The atmosphere contains only 1 % of the total global carbon pool. The undesirable changes can be reversed by increasing soil and eco system carbon pools, increasing mean resident time(MRT) of sequestered carbon and decreasing GHG emission to atmosphere.

Contribution of livestock to GHG emission

The livestock sector uses one third of global arable land. It accounts for 8% of global freshwater resources. It is also largely responsible for deforestation and biodiversity loss. In 2015, global GHG emissions reached 57 gigatonnes (Gt), with livestock accounting for 15% of it. The meat sector alone produced over 5 Gt of emissions during the same year. About 13 billion tonnes of manure produced per year requires safe disposal.

Herreroa *et al.*(2013) analyzed information on biomass use, production, feed efficiency, excretion, and greenhouse gas emissions for 28 regions, 8 livestock production systems, 4 animal species (cattle, small ruminants, pigs, and poultry), and 3 livestock products (milk, meat, and eggs). The total non-CO_2 GHG emissions from the livestock sector in 2000 were 2.45 Gt CO_2 eq. Methane from enteric fermentation from ruminants, was by far the largest source of GHG emissions (1.6 Gt CO2eq). Methane and nitrous oxide (N_2O) from manure management, and N_2O from manure application to soils were 0.25, 0.21, and 0.49 Gt CO2eq, respectively. Cattle accounted for 77% of emissions. The contribution of monogastrics to GHG emissions was only 10% of total livestock emissions, and most of this is in the form of methane from manure management (56% of their total emissions).The developing world contributes 75% of global GHG emissions from ruminants and 56% of emissions from monogastrics. Mixed crop–livestock systems produce the bulk of emissions from ruminants (61%), and grazing systems account for 12% of emissions.

Greenhouse gas emissions from Indian livestock

Livestock constitutes an integral component of Indian agriculture sector and also a major source of GHGs emissions. The total methane emission including enteric fermentation and manure management of livestock was estimated at 11.75 Tg/year for the year 2003. Enteric fermentation constitutes ~91 % of the total methane emissions from Indian livestock. Dairy buffalo and indigenous dairy cattle together contribute 60 % of the methane emissions. The total nitrous oxide emission from Indian livestock for the year 2003 is estimated at 1.42 Gg/year, with 86.1 % contribution from poultry. The total GHGs emission from Indian livestock is estimated at 247.2 Mt in terms of CO2. Using the remote sensing derived potential feed/fodder area available for livestock, the average methane flux was calculated as 74.4 kg/ha (Chhabra *et al.,* 2013).

Types of GHG from animal agriculture

The most important GHGs from animal agriculture are methane and nitrous oxide. These are among the more potent GHGs: one molecule of methane will trap 21 times more heat than carbon dioxide, while one molecule of nitrous oxide will trap 310 times more heat than carbon dioxide.

Methane

Enteric fermentation is the fermentation that takes place in the digestive systems of animals. In particular, ruminant animals (cattle, buffalo, sheep goats and camels) have a large "fore-stomach," or rumen, within which microbial fermentation breaks down food into soluble products that can be utilized by the animal. The microbial fermentation that occurs in the rumen enables ruminant animals to digest coarser plant material than monogastric animals. Methane is produced in the rumen by bacteria as a byproduct of the fermentation process. This CH_4 is exhaled or belched by the animal and accounts for the major portion of emissions from ruminants. Methane also is produced in the large intestines of ruminants and is expelled. In this first stage of digestion, the forage is acted on by the varied population of microorganisms, including bacteria, fungi and protozoa in the fore-stomach. This process releases hydrogen, while producing volatile fatty acids and microbial cells containing energy and essential proteins to be made available for the growth of the animal. In ruminants, the hydrogen is removed through the action of a group of microbes called methanogens that gain their energy through combining carbon dioxide with hydrogen to form methane. There are a variety of factors that affect methane production in ruminant animals, such as the feed characteristics, the feeding level and schedule, the use of feed additives to promote production efficiency and the activity and health of the animal (Thampi and Thomas, 2013).

Methane accounts for 33 percent of the total GHG emissions in agriculture and 71 percent of all agricultural sources of methane. Activities related to the storage and land application of manure release 12 percent of the total agricultural methane emissions, and represent 25 percent of all agricultural sources of methane.

Nitrous Oxide

Nitrous oxide is released in soils after the application of synthetic and organic fertilizers including manures.

Climate change management

Climate change management strategies are classified into two categories viz. mitigation and adaptation strategies.

Mitigation

The term mitigation refers to activities which reduce emission of green house gases(GHGs) by anthropogenic activities and enhances carbon sinks through natural and engineered processes. These include preventive measures to reduce GHG emissions.

Adaptation

The term adaptation refers to activities which reduce risks of productivity decline because of global warming and climate change. These are curative measures to sustain crop and animal productivity.

Strategies to check GHG emission from livestock sector

By 2050, the Earth's population is expected to be more than 9 billion. The need for a secure food and water supply will force agriculture to increase production in a time of climate uncertainty. Like all economic sectors, animal sector should reduce carbon fruit print in an economically viable manner.

The strategies can be discussed under four heads.

1. Animal husbandry (animal breeding, feed supplements, improved pastures)
2. Management systems (stocking rates, biological control)
3. SMART waste management
4. Numbers of livestock
5. Increased animal welfare
6. Better education

A) Animal husbandry

i) Animal breeding: There are variations among animals in methane emissions per unit of feed intake and these variations suggest that there may be heritable differences in methanogenesis (methane production). Trials suggest that animal breeding could achieve a 10–20% reduction in methane emissions. While breeding for reduced methanogenesis may not be compatible with other breeding objectives, breeding for improved feed conversion efficiency (lower net feed intake) should be compatible and is likely to reduce methane emissions and the greenhouse gas intensity of animal products.

ii) Diet supplements and feed alternatives: Supplements include oils, fats, tannins, probiotics, nitrates, enzymes, marine algae and Australian native vegetation. Methane abatements of 10–25% are possible by feeding ruminants dietary oils, 37–52% abatement achieved in individual studies. Plant secondary compounds, such as condensed tannins, have been shown to reduce methane production by 13–16%, mainly through a direct toxic effect on methanogens. However, high condensed tanins concentrations can reduce voluntary feed intake and digestibility. Plant saponins (natural steroids occurring in several plant families) also potentially reduce methane, with methane suppression attributed to combating protozoal infections.

Role of biochar in reducing GHG emission

Charcoal has been used to treat digestive disorder in animals since several thousand years. But only since 2010, biochar has increasingly been used as regular feed additive in animal farming usually mixed with standard feed at approximately 1% of the daily feed intake. The use of biochar as feed additive has the potential to improve animal health, feed efficiency and the animal-stable environment; to reduce nutrient losses and GHG emissions and to increase soil organic matter and thus soil fertility. Feeding biochar(activated) for all investigated livestock species has positive effects on parameters like toxin adsorption, digestion, blood values, feed use efficiency and livestock weight gain, meat quality and GHG emissions. The energy efficient digestion and thus higher feed efficiency are due to direct electron transfer between different species of bacteria or microbial consortia via the biochar mediator in the animal digestion tract. It results in selective probiotic effect, reduced N-losses and reduced GHG emissions. The poultry birds, pigs, fish and other omnivore animals provoke GHG emissions (mainly NH_3, CH_4, N_2O) when their liquid and solid excretions decompose anaerobically. But ruminants cause direct methane emissions through flatulence and burps (eructation). It is likely that microbial decomposition of manure containing digested biochar produces less ammonia, less methane and thus retain more nitrogen, as evident from the study of composting manure with and without biochar or when biochar is used as bedding or additive in manure. Feeding 0.3 and 1% biochar could replace antibiotic treatment in chicken and ducks, respectively. Feeding biochar could thus have an indirecteffect on GHG emissions when it is able to replace regular antibiotic "feeding" that produces high indirect GHG emissions after soil application of antibiotic contaminated manure. Feeding biochar to grazing cows had positive secondary effects on soil fertility and fertilizer efficiency reducing mineral N-fertilizing requirements which could be another indirect biochar GHG mitigation effect. If the global livestock would receive 1% of their feed in form of biochar, a total of about 400 Mt of CO2 eq or 1.2 % of the global CO_2 emissions could be compensated. Biochar feeding improves animal health and nutrient efficiency, reduces enteric methane emissions, minimizes GHG emissions from manure management and for sequesters carbonin soil with fertility improvements(Schmidt *et al.*, 2017).

iii) Improved pastures: Improved forage quality with lower fibre and higher soluble carbohydrates can reduce methane production in livestock. Being structural fibres, cellulose and hemi-celluloses ferment more slowly than non-structural carbohydrates and yield more methane per unit of feed digested. Methane emissions are commonly lower with more forage legumes in the diet, partly because of the lower fibre content (faster rate of digestion) and in some cases, the presence of condensed tannins. As improved diet increases animal growth and reduces methane production, it has the effect of reducing the greenhouse gas intensity of the animal products. Pasture quality can be improved in several ways including by plant breeding, changing from tropical

(C_4) to temperate (C_3) grasses that use different pathways to capture carbon dioxide, or grazing on less mature pastures. Several alternative plant forages, such as broccoli leaves and some native plants such as *Eremophila glabra*, *Acacia saligna* and a number of saltbush species, have been shown to reduce methane emissions in laboratory experiments

2) Management systems

i) Stocking rates: Reducing the number of unproductive animals on a farm can potentially improve profitability and reduce greenhouse gas emissions. If productivity increases through nutritional and breeding strategies, the number of livestock can be reduced without losing the quantity of meat that is currently produced. Strategies such as extended lactation in dairying , where cows calve every 18 months rather than annually, reduce herd energy demand by 10%, and so potentially reduce methane emissions by a similar amount.

ii) Biological control: Three biological control methods are being tested for their ability to reduce methane production from livestock, using viruses (to attack the microbes which produce methane), specialised proteins (to target methane-producing microbes) and other microbes (methanotrophs) to break down the methane produced in the rumen into other substances.

3) Smart waste management

i) Decrease in meat waste: About 20% of all meat goes to waste every year. Although most waste is generated at the consumer end, the root causes lie across the whole supply chain, e.g. poor packaging or bad meat quality. Smart waste management offers high potential to first identify such root causes and second, to improve existing processes and thereby decrease waste amounts.

ii) On-farm energy generation: Biogas generation and anaerobic digestion technologies should be adopted.

iii) Vermicomposting and composting

iv) Aerobic oxidation

4. Number of livestock

i) Animal productivity enhancement: Fewer animals are needed to produce the same amount of product by promoting more output per unit input such as meat, milk, and eggs. Hence, the GHG emission will be less.

5. Increased animal welfare

6. Better education

Impact of climate change on animal agriculture

Effect of heat waves on mortality of animals

Significant heat events since the mid 1990s appear to be increasing and have resulted in sizable human and animal mortality. Over 800 peacocks died during a heat wave in India in 2007(Gaughan *et al.,* 2009).

Impact of climate change on animal health

The foremost reaction of animals under thermal weather is increases in respiration rate, rectal temperature and heart rate. It directly affect feed intake thereby, reduces growth rate, milk yield, reproductive performance, and even death in extreme cases. Dairy breeds are typically more sensitive to HS than meat breeds, and higher producing animals are, furthermore, susceptible since they generates more metabolic heat. HS suppresses the immune and endocrine system thereby enhances susceptibility of an animal to various diseases. Hence, sustainable dairy farming remains a vast challenge in these changing climatic conditions globally. The foremost reaction of animals under thermal weather is increases in respiration rate, rectal temperature and heart rate. It directly affect feed intake thereby, reduces growth rate, milk yield, reproductive performance, and even death in extreme cases. Dairy breeds are typically more sensitive to HS than meat breeds, and higher producing animals are, furthermore, susceptible since they generates more metabolic heat. HS suppresses the immune and endocrine system thereby enhances susceptibility of an animal to various diseases. Hence, sustainable dairy farming remains a vast challenge in these changing climatic conditions globally (Das *et al.,* 2016).

Climate changes can affect the health of livestock and poultry, both directly and indirectly. Direct impacts include temperature-related illness and death, and the morbidity of animals during extreme weather events. Indirect impacts follow more intricate pathways and include those deriving from the influence of climate on microbial density and distribution, distribution of vector-borne diseases, host resistance to infections, food and water shortages, or food-borne diseases. Cooks *et al.*(2002) reported higher occurrence of mastitis during periods of hot weather in dairy animals. During summer, ketosis is more prevalent due to increased maintenance requirements for thermoregulation and lower feed intake (Lacetera *et al.*,1996), and the incidence of lameness increases as a consequence of metabolic acidosis(Shearer, 1999). Bernabucci *et al.*(2002) reported alteration of liver function, mineral metabolism and oxidative status in dairy cows.

Animal Adaptation to climate change

Adaptation

Adaptation is the appearance or behaviour or structure or mode of life of an organism that allows it to survive in a particular environment". Adaptations can be observed in structure or behaviour or physiology of an organism. Adaptations have a genetic basis and have been evolved and perfected through the evolutionary process.

Acclimation

Acclimation (or hardening) is the non-heritable modification of organism characters caused by exposure to new environmental conditions such as warmer or drier weather. It results from temporary modifications to the organism phenotype caused by the changing environment.

Animal management for adaptation to climate change

1. Genetic Modifications
2. Environmental Modifications
3. Nutritional Modification

Genetic Modifications

Phenotypes within the breed types having high heat tolerance are preferred. Similarly phenotypes within the heat tolerant breeds are selected. Emphasis is given to identify breeds that meets market requirements and simultaneously heat tolerant. Cattle breed with shorter hair and lighter colour have "slick" gene and are heat tolerant.

Environmental Modifications

1. Provision of shade
2. Provision of air movement/ventilation
3. Provision of water source nearby for cooling of livestock
4. Direct water application for cooling

Nutritional Modification

A 'cold' diet for dairy cow generates a high net nutrient proportion for milk production and lower heat increment. The 'cold' diets have higher energy contents per unit volume, highly fermentable fiber, lower protein degradability and high by-pass nutrients contents.

Smart fodder production

Forages as a means to mitigate GHG Emissions

Agricultural production systems including improved forages can address mitigation of all major GHG. Forages mitigate GHG emissions in three ways: by sequestering atmospheric CO2; by reducing ruminant CH_4 emissions per unit livestock product as compared to a lower quality rangeland/degradedpasture; and by reducing N_2O emissions.

Sequestering atmospheric CO2

It is estimated that the soil organic carbon sequestration potential of the world's grasslands was 0·01 to 0·3 Gt C/year.

Reducing ruminant CH_4 emissions per unit livestock

There are different strategies to reduce methane emissions such as forage diets with high digestibility plus high energy and protein concentrations, inclusion of forage legumes in diet and use of forages in mixed crop-livestock systems. Forage diets with high digestibility plus high-energy and protein concentrations produce less CH_4 per unit of meat or milk produced. Forage and feed with a high proportion of easily digested carbohydrates such as starches and sugars usually move through the rumen faster and are used more efficiently than forage and feed with a high proportion of roughage such as cellulose. Grain has a higher proportion of easily digested carbohydrates, especially starch, than forage, and is therefore used more efficiently. Legumes contain less structural carbohydrates and more condensed tannins than does grass, and adding legumes to the diet can further reduce CH_4 emissions per unit of meat or milk produced. Methane emissions are also commonly lower with higher proportions of forage legumes in the diet, partly due to lower fibre content, faster rate of passage and, in some cases, the presence of condensed tannins. Another such approach is the use of tannin containing forages and breeding of forage species with enhanced tannin content.

Reducing nitrous oxide emissions

Some plants release biological nitrification inhibitors (BNIs) from their roots, which suppress nitrifier activity and reduce soil nitrification and N_2O emission. This biological nitrification inhibition (BNI) is triggered by ammonium (NH4+) in the rhizosphere. The release of the BNIs is directed at the soil microsites where NH4+is present and the nitrifier population is concentrated. Tropical forage grasses, cereals and crop legumes show a wide range in BNI ability. The tropical Brachiaria spp. have high BNI capacity, particularly *Brachiaria humidicola* and *Brachiaria decumbens*. Nitrification inhibitor in *Brachiaria* grasses is brachialactone,a cyclicditerpene. Brachiaria pastures can suppress N_2O emissions.

Increasing the efficiency of N use in the ruminant animal reduce nitrous oxide emissions from ruminants. Rapid breakdown of herbage proteins in the rumen and inefficient incorporation of herbage nitrogen by the rumen microbial population are major causes of N loss and gaseous emissions. When sheep and cattle are given fresh forages they can waste 25-40 per cent of forage protein. Genetic improvement of the forage grasses and legumes that constitute important components of the ruminant diet has the potential to reduce emissions toair.

Two possible strategies for increasing the efficiency of conversion of forage-N to microbial-N have been suggested:

1. Increasing the amount of readily available energy accessible during the early part of the fermentation
2. Providing a level of protection the forage proteins, thereby reducing the rate at which their breakdown products are made available to the colonising microbial population.

One approach is to develop forage species with a better balance between water soluble carbohydrate (WSC)and crude protein (CP) by increasing the WSC content of the grass or the clover component or reducing the protein content of the legume.

The most advanced of these approaches is the development at IGER of high WSC rye grasses where more N is partitioned into meat and milk and less is available for nitrogenous emissions through excreta (Miller *et al*., 2001).

Another approach is increasing the content of compounds that affect protein breakdown in the rumen. Opportunities also exist within forages to select for other specific traits that can reduce protein loss.

A good example of this approach is the emerging research on the enzyme polyphenol oxidase (PPO),which is at a particularly high level of activity in red clover in comparison with other species and has a role in protein protection. This enzyme converts phenols to quinones which subsequently bind to protein and slow the rate of protein degradation. Thus, in silo, the protein made available for diffuse pollution of nitrogen e. g. as ammonia, is reduced. Ensiling alfalfa (lucerne) leads to the degradation of44 to 87 per cent of forage protein to non-forage protein (NPN). In comparison, red clover has up to 90per cent less protein breakdown. Increasing the level of PPO is a target for genetic improvement in red clover as a route to reduced nitrogenous pollution (Thampi and Thomas, 2017).

Forage crop production schedule for smart animal agriculture

Crop	Scientific name	Family	Growth cycle and time of sowing/ planting
Non-leguminous forages			
Jowar(chari)	Sorghum bicolor(L.) Moench	Poaceae	Annual, March to July (North), February to November(South)
Bajra	Pennisetum americanum L. (P. glaucum)	Poaceae	Annual, April to July
Napier xbajra hybrid	Pennisetum purpureumxP. americanum (P. purpureumx P typhoides)	Poaceae	Perrenial, March to October(North) Entire year(South)
Maize	Zea mays L.	Poaceae	Annual, April to August(North), February to November(South)
Leguminous forages			
Cowpea	Vigna unguiculata (L.) walp	Fabaceae	Annual, March to july
Clusterbean	Cyamopsis tetragonolaba (L.) Taub.	Fabaceae	Annual, April to may as summer crop with irrigation, June to August as kharif crop, Planting in June causes rank vegetative growth
Ricebean or red bean	Vigna umbellate (Thunb.) Ohwi & Ohashi	Fabaceae	Annulal, June to first week September as rainfed crop, February onwards as irrigated crop

Rabi forage crops

Non-leguminous forages			
Jaie of Oats	Avena sativa L	Poaceae	Annual, October – November
Jau or Barley	Hordeum vulgare L. Emond. Bowden	Poaceae	Annual, October-November
Chara sarson or Chinese cabbage	Brassica pekinensis (Lour.) Rupr	Brassicaceae	Annual, September-October
Suljam or turnip	Brassica rapa L.	Brassicaceae	Annual,m September to mid october

Leguminous forages			
Berseem or Egyptian clover	Trifolium alexandrinum L.	Fabaceae	Annual, October-november
Lucerne or Alfa alfa	Medicago sativa L.	fabaceae	Annual, September to Mid November
Senji or sweet clover	Melilotus indica L.	Fabaceae	Annual, October-November

Role of digitalization in smart livestock farming

The smart livestock farming should be sustainable, measurable, achievable, replicable with tangible gains. Various aspects of digitalization viz.Artificial Intelligence (AI), Information and Communication Technologies (ICT), Cyber-Physical Systems (CPS) and Internet of Things (IoT), cloud computing, data analytics, platforms and crowd sourcing will facilitate Smart livestock farming.

References

Bernabucci, U, Ronchi, B, Lacetera, N, Nardone, A.2002. Markers of oxidative status in plasma and erythrocytes of transition dairy cows during the hot season. Journal of Dairy Science 85:2173–2179

Chhabra, A., Manjunath, K. R., Panigrahy, S. and Parihar, J. S. 2013. Greenhouse gas emissions from Indian livestock. Climatic Change 117:329 344

Cook, N.B,, Bennett, T.B, Emery, K.M, Nordlund, K.V. (2002) Monitoring nonlactating cow intra mammary infection dynamics using DHI somatic cell count data. Journal of Dairy Science 85:1119–1126

Das, R., Sailo, L., Verma, N., Bharti, P., saikia, J., Imtiwati and Kumar, R. 2016. Impact of heat stress on health and performance of dairy animals: A Review Veterinary World. 9(3): 260–268.

Gaughan, J., Lacetera, N., Valtorta, S. E., Khalifa, H.H., Hahn, L. and Mader, T..2009. Response of domestic animals to climate challenges (pp. 131-170, Chapter 7). In book : Biometeorology for Adaptation to Climate Variability and Change

Herreroa, M., Havlík, P. and Valin, H. Notenbaert, A., Rufino, M. C., Thornton, P. K., Blümmel, M., Weiss, F. Grace, D., and Obersteine, M. 2013. Biomass use, production, feed efficiencies, and greenhouse gas emissions from global livestock systems. PNAS 110(52) : 20888 -20893

Lacetera, N, Bernabucci, U, Ronchi, B, Nardone, A.1996. Body condition score, metabolic status and milk production of early lactating dairy cows exposed to warm environment. Riv Agr Subtrop Trop 90:43–55

Shearer JK .1999. Foot health from a veterinarian's perspective. Proceedings Feed and Nutritional Management Cow College, Virginia Tech, pp. 33–43

Singh, B., Kumar, S., Dhaka, A. K. and Kumar, S. 2011. Intercropping of cereals and legumes for forage production in kharif season–a review. Forage Research 36 : 189-196.

Somashekar, K. S., Shekara, B. G., Kalyanamurthy, K. N. and Lalitha, B. S.. 2014. Green fodder augmentation for increasing animal productivity–a review. Forage Research 40 (3) : 147-153.

Thampi, A. C. and Thomas, U. C. 2017. Mitigation of climate change through fodder production systems–a review. Forage Research 43 (2) : 73-80

Trupa, A., Aplocina, E. and Degola, L.. 2015. Forage quality and feed intake effect on methane emissions from dairy farming. Engineering for Rural Development 601-605.

18

Startegies for the Conservation of Livestock Breeds in Agro-climatic Zones of Tamil Nadu

S Vasantha Kumar

Livestock farming is an important activity to a large population of small and marginal farmers as well as landless agricultural labourers of Tamil Nadu. It also provides employment opportunities for most of the unemployed and underemployed in the rural areas. Tamil Nadu is bounded by the Eastern Ghats on the north, by the Nilgiris, the Anamalai Hills and Kerala on the west, by the Bay of Bengal in the east, and by the Indian Ocean on the south. Tamil Nadu is the eleventh largest state in India by area. The land area has been classified into seven agro-climatic zones based on soil characteristics, rainfall distribution, irrigation pattern, cropping pattern and other ecological and social characteristics. Tamil Nadu endowed with 13.97 per cent of recognised breeds of the domestic ruminants (cattle, buffalo, sheep and goats) with 22.55 millions of population (19thLivestock Census-2012).

Table1: Agro-climatic zones of Tamil Nadu

S. No.	Agro-climatic zones
1.	North Eastern
2.	North Western
3.	Western
4.	Cauvery delta
5.	Southern
6.	High rain fall
7.	Hill and high altitude

Cross breeding should be implemented only in potential areas having availability of quality feeds at relatively low prices, demand for cow milk and low demand for draught animals and accessibility to veterinary facilities. Only the non-descript animals should be bred through cross breeding. In Tamil Nadu, there are many organisations have been involved for the establishment of livestock farms for the conservation of draught breeds of cattle, meat and wool breeds of sheep and meat breeds of goats in their home tracts.

Table 2: Distribution of livestock breeds in Agro-climatic zones of Tamil Nadu

Livestock Species	Livestock Breeds	Distribution in agro-climatic zones
Cattle	Kangeyam	Western
	Bargur	Western
	Umblacherry	Cauvery delta
	Pulikulam	Southern
Buffalo	Toda	Hill and high altitude
	Bargur	Western
Sheep	Madras red	North Eastern
	Mecheri	North Western
	Coimbatore	Western
	Trichi black	Cauvery delta
	Nilgiri	Hill and high altitude
	Vembur	Southern
	Kilakarsal	Southern
	Ramnad white	Southern
	Chevad	Southern
	Kachaikatti	Southern
Goat	Salem Black	North Western
	Kanni adu	Southern
	Kodi adu	Southern

Table 3: Distribution of livestock population in Tamil Nadu

Species	India (millions)	Tamil Nadu (millions)	Distribution (per cent)
Cattle	190.9	8.81	4.61
Buffalo	108.7	0.78	0.72
Sheep	65.06	4.78	7.35
Goat	135.17	8.18	6.05
Total	499.83	22.55	18.73

Table 4: Recognised livestock breeds of Tamil Nadu

Species	No. of recognized breeds		Distribution (Per cent)
	India	Tamil Nadu	
Cattle	43	04	9.30
Buffalo	16	02	12.50

Sheep	43	10	23.26
Goat	34	03	8.82
Total	136	19	13.97

1. District Livestock Farms

For the production of superior germplasam of recognised breeds of cattle, sheep and goats, the Department of Animal Husbandry, Govt. of Tamil Nadu established livestock farms in different agro-climatic regions of the State. In District Livestock Farm at Korukkai a Umblacheri cattle breed is being conserved for the benefit of farmers. They also established Sheep farms at Sattur (Virudhunagar dist), Mugundarayapuarm (Vellore dist) China Salem (Salem dist), and other District Livestock Farms in various districts under the control of Department of Animal husbandry distribute the better rams for breeding to the private growers.

2. Tamil Nadu Cooperative Milk Producers' Federation

Milk recording in Kangeyam cattle was implemented by Tamil Nadu Cooperative Milk Producers' Federation with the financial assistance from the Tamil Nadu Livestock Development Agency. From the elite Kangeyam cows that are enrolled in the programme, high yielding cows selected are inseminated and the bull calves born to these cows are selected and reared for semen production in the semen stations. The programme is implemented in the milk unions of Coimbatore, Erode and Tiruchirapalli.

3. University Research Farms

Tamil Nadu Veterinary and Animal Sciences University (TANUVAS) research farms in different locations in Tamil Nadu also contributed for the improvement of livestock breeding and conservation. For conservation Kangeyam, Bargur and Pulikulam cattle there are new research stations have been established to preserve and produce the germplasm for the benefit of the farmers. On the subject of small ruminant production, research farms also established to conserve the Madras Red, Mecheri, Kilakarsal, Ramnad white, Nilgiri, Vembur and Chevad breeds of sheep and Kanni, Kodi and Salem black breeds of goats in different agro-climatic region of Tamil Nadu.

4. Non Government Organisations

Senaapathy Kangeyam Cattle Research Foundation (SKCRF)

Senaapathy Kangayam Cattle Research Foundation (SKCRF) is an in situ conservation and breeding centre based in Kuttapalayam in Tirupur district of Tamil Nadu. SKCRF functions as a resource and research centre working on conservation of native breeds of cattle. Awareness is focussed more intensely for the Kangayam breed and the Korangadu, a unique silvi pastoral land and farming system which is unique to Kongu region, mid-western Tamilnadu. SKCRF also works to preserve the bio-cultural values and the uniqueness around our indigenous breeds of livestock and the traditional knowledge associated with it. Thiru. Senaapathy is the eighth generation breeder of Kangayam cattle in his family and he runs the Senaapathy Kangayam Cattle Research Foundation.

Sustainable Agriculture & Environment Voluntary Action (SEVA)

A Nongovernmental Organisation SEVA functioning on empowers marginalized communities through traditional knowledge, grassroots innovations and conservation of agricultural biodiversity.

SEVA has been involved in the following activities

1. Documentation and dissemination of indigenous knowledge and grassroots innovations and conservation of local livestock breeds in India.
2. Organising training programme on herbal preparations for preventive and curative treatment of livestock diseases.
3. Capacity building of innovators, pastoralists, livestock keepers and farmers for testing, demonstration and dissemination of their products, knowledge or breeds conserved by them.

SEVA has conducted awareness raising through village level training cum workshops, animal health camps using herbal preparations, breeders group meeting micro credit for members. Regular deworming, preventive treatment and providing grazing rights in forests will encourage the local herders for up keep of sheep population. SEVA has promoted Sheep Breeders Association and Breed Saviour Award 2010 has been given to the association.

Management practices for the conservation of livestock

Korangadu pasture land

It is a traditional dry land grass farming system to earn Income in dry lands. "Korangadu" is a traditional pastureland farming system existing in semi arid tract of Tamil Nadu. Korangadu is typically a mixure of grass, legumes and tree species including annual and perennials. Approximately 50,000 ha. of Korangadu pasture land is noticed in 500 villages in Erode, Karur, Dindigul and Coimbatore Districts of Tamil Nadu State, Southern India. The size of individual paddocks of Korangadu land ranges from 1.5 ha to 10 ha depending upon the wealth status or ownership pattern by farmers. Farmers or landless livestock holders keep sheep, cattle, buffalo and Korangadu provides baseline of livelihoods for them by feeding their animals. This region receives annual rainfall of 600 - 675 mm. The soil is laterite red soil or with gravel type and water will not stagnate on any amount of rainfall. The region situates in rain shadow region of Western ghats. The majority of the rural population depends upon livestock; they are settled agro pastoralsits and allow their animals to graze in their own grassland paddocks confined to 2 - 4 ha size. Korangadu has predominantly 3 major species of flora which are spatially in 3 tiers. The lower tier is grown with grass Cenchrus ; tree species include *Acacia leucophloea* locally called as Velvel and land is fenced with thorny shrub locally called as Kiluvai (*Commiphora berryii*) as live fence.

Penning system

Penning is a practice in which the animals were kept in the agricultural land of farmers for overnight for a stipulated period based on their requirement to enrich the field with manure. The penning charges were slightly variable as per the demand of it. The penning duration was generally 6 PM to 9 AM of next day i.e. for 14–15 h. When there was no demand for penning, the herds were kept in any wasteland area where the dung was collected, dried and sold in the market.

The sheep herds of individual herders herd numbering 450 animals are penned during night time for manuring the farmers field. For one acre field can be manured in 2 nights. A group of 20 sheep constitute one 'Moi'. If 60 sheep are maintained by a herder then it will be 3 'Moi'. In a month there will be about 24 Mois. If daily penning income is Rs.310 for the entire herd then a person with 60 sheep (3 Mois) will receive 3 days income (Rs.1830) continuously. After a cycle of 24 Mois it will start again fresh. A person with 60 sheep will receive a monthly income of Rs.1800 per cycle (24 days) through penning. In Pulikulam cattle the penning is also a practice in which they charges @`400 to 500 per night for the herd of about 400-500 animals, there by getting about Re 1/animal/night.

Bull riding (Jellicut)

Pulikulam, which is maintained as big migratory herds in the Madurai area of Tamil Nadu. As informed by the herders of the area, the origin of the Pulikulam cattle is Pulikulam village of Sivagangai district of Tamil Nadu. This breed is famous for Jellicut and bullocks are used for all kind of agricultural operations and rural transport. The breeding tract of Pulikulam cattle lies between

9°30'and 10°30'N latitude and between 77°47' and 78°49'Elongitude. The climate of the region is semi-arid tropical and has higher mean temperature and low degree of relative humidity. Virudhunagar and Sivagangai districtsof Tamil Nadu.The number of herds per village varied from 1 to 24.The breed is also famous for the local game called Jellicut, meaning an ornament of leaves, during Pongal festival inTamil Nadu. In this event, the horn of the bull is decorated with vividly coloured cloth and sometimes even ornaments. The cloth or ornaments in the cloth is firmly tied round the horn of bull and it is set free. Many persons attack on the bull to untie the cloth, bull becomes excited and fights. The bull is controlled and thrown down before the cloth is untied. There are chances that person may get severe injury or even die but the event is very popular. Their calves are usually sold to be grown as draught animal, for ploughing and for Jallikattu (bull riding) during harvesting festival occasion (Jan-Feb).

References

Ganapathi,P., Rajendran, R and Subramanian, A. Distribution and Population status of Bargur cattle, Indian Veterinary Journal, 2009;86: 971-972.

Livestock 19th census, 2014. Department of Animal Husbandry and Dairying, Ministry of Agriculture, Government of India, New Delhi, India.

Pundir, R.K.; Kathiravan, P. Singh, P.K. Manikhandan, V. A. Bargur, the hill cattle of Tamil Nadu Indian Journal of Animal Sciences 2009. Vol.79 No.7 pp.681-685.

Singh, P.K., Pundir, R.K., Kumarasamy, P and Vivekanandan, P. Management and physical features of migratory Pulikulam cattle of Tamil Nadu. Indian Journal of Animal Sciences, 82 (12): 1587–1590, December 2012.

Thesinguraja. S., Mathialagan P., Thilakar P., Devendran P and Palanichamy, V. Socio-Economic Profile of Pulikulam Cattle Rearers in Madurai and Sivagangai Districts of Tamil Nadu, India Int.J.Curr.Microbiol.App.Sci., 2017. 6(12): 424-429.

19

Sustainability of Climate Resilient Livestock Conservation Strategies in India

S K Dash

Introduction

Climate change represents a threat not only to the existence of individual species, but also to the genetic diversity hidden within them. The finding promises to complicate assessments of how climate change will affect biodiversity, as well as conservationists' task in preserving it.

Climate change comes as an additional factor affecting a livestock sector that is already highly dynamic and facing many challenges. Important objectives of AnGR management include ensuring that AnGR are effectively deployed to meet these challenges (i.e. are well matched to the production environments in which they are kept) and that the genetic diversity needed to adapt production systems to future changes is maintained. Climate change is likely to create a number of problems in many areas of animal husbandry (housing, feeding, health care, etc.) and threaten the sustainability of many livestock production systems and their associated AnGR. At the same time, many of the specific challenges associated with climate change (high temperatures, disruptions to feed supplies, disease outbreaks, etc,) as well as the general unpredictability it brings to the future of the livestock sector, highlight the importance of retaining diverse genetic options for the future. The upper critical temperature of dairy cattle is lower than other livestock species (Wathes *et al.*,1983).

Climate change is predicted to lead to a rise in temperatures globally. In parts of the world it is likely that this will increase the problem of heat stress in livestock and require adjustments to husbandry and production strategies. Although the high-output breeds that increasingly dominate global livestock production are usually not well adapted to heat stress, the high external input systems in which they are often kept provide several options for alleviating the effects of high temperatures, including the use of cooling systems (sprays, etc.)

and adjusting the diet to reduce metabolic heat production. Such production systems are generally quite well able to protect animals from the local-scale effects of climate change. However, their heavy reliance on external inputs may make them vulnerable to rising prices – a problem that may be exacerbated by climate change. In production systems where heavy use of external inputs is not possible, the importance of well adapted animals is likely to increase under climate change.

Underneath factors on diversity

Animals do have specific responses to climate change. Species respond to climate changes by migration, adaptation, or if neither of those occur, death. These migrations can sometimes follow an animal's preferred temperature, elevation, soil, etc., as said terrain moves due to climate change. Adaptation can be either genetic or phenological and death can occur in a local population only (extirpation) or as an entire species, otherwise known as extinction.

Every organism has a unique set of preferences or requirements, a niche and biodiversity has been tied to the diversity of animals niches. These can include or be affected by temperature, aridity, resource availability, habitat requirements, enemies, soil characteristics, competitors and pollinators. Since the factors that compose a niche can be so complex and interconnected, the niches of many animals are bound to be affected by climate change.

Global climate change may lead to the loss of significant amounts of hidden diversity, even if some of the traditionally defined species will persist. The most genetically diverse regions are also the most endangered. This combination of genetic diversity and vulnerability has been found for all the species. Genetic diversity is gaining increased attention in conservation circles. Through our work to determine climate-adaptation strategies, we realize that genetics is one way to get an overall better view of how species are affected by climate change. Combining genetics and ecology should aid conservation efforts to get sustainable result.

Adaptability

The concept of adaptability revolves around fitness describing relative ability of an individual to survive and reproduce next generation to ensure continued survival of the population and is the result of natural selection over many generations. Current trend in genetic selection has severely eroded the genetic base ignoring the diversity, importance of adaptation, production of multiple products, social and cultural value of the livestock. Unplanned genetic introgression and crossbreeding has contributed to the greatest extent toward the loss of indigenous breeds. The genetic mechanism influencing fitness and adaptation is not well explored and adaptation traits are usually characterized by low heritability. Further, it may be difficult to combine the adaptation traits with high production potential as there seem to be different physiological and metabolic processes involved. Though decision regarding matching genotypes with environment or vice versa will be situation specific, the low and intermediate level of animal production in many parts of the world suggests that increased yields and efficiency will be more environmentally sustainable than extensive goals ensuring genetic diversity, environmental soundness, animal health and welfare and social viability. Breeding for climate change adaptation or mitigation will not be necessarily different from existing programs. Breeding indices should be balanced to include traits associated with heat resilience, fertility, feed conversion efficiency, disease tolerance and longevity in addition to higher productivity and give more consideration to genotype by environment interactions (GxE) to identify animals most adapted to specific conditions and natural stratification of breeds and species across climatic zones.

To illustrate the case of variation in adaptability, there are numerous examples like zebu cattle that is uniquely suited to hot and humid climates because of its smooth coat, primary hair follicles, improved sweat and sebaceous glands and better ability to lose moisture by evaporation than Bos taurus cattle (Turner, 1980). Studies have shown milk yield of crossbred cows in India (e.g., Karan Fries, Karan Swiss and other Holstein and Jersey crosses) to be negatively correlated with temperature-humidity index (Shinde *et al.* 1990; Kulkarni *et al.* 1998; Mandal *et al.* 2002)

Environmental factors such as temperature, humidity, wind velocity and radiation determine comfort and stress levels of animals. Genetic adaptation to adverse environmental conditions including heat stress is a slow process and is the result of natural selection over many generations (Ames and Ray, 1983).

Genetic Manipulation in Livestock Production

Exploiting genetic variation to breed superior animals is often misunderstood and sometimes thought to be panacea to solve all problems associated with unfavorable environments. Genetic adaptation is relatively a slow process which is subject to ever-changing geological, biological and climatic conditions (Dobzhansky, 1970).

A dichotomy exists in the sense that genetic adaptation and fitness are synonymous in terms of evolution, implying that only those who survive and reproduce for next generations in a given set of environmental challenges are fit and thus better adapted. A given breed of livestock species may have evolved over many generations for particular agro-ecological system and this knowledge is often ignored due to excessive focus on short-term strategies and benefits.

Genetics and Adaptive Traits for Harsh Environment

Evolution is a continuous process which molds and remolds combinations of genes in living things to adapt to changing environment. A trait is an aspect of the developmental pattern of the organism. An adaptive trait, therefore, is an aspect of the developmental pattern which facilitates the survival and/or reproduction in animals in a certain succession of environments (Dobzhansky 1970). Fitness is an important trait for any species which centers on the adaptation of individual species in a given environment to propagate their genes in subsequent generations. Livestock keepers throughout the world have been practicing animal husbandry and breeding in unfavorable environment for centuries. As a result, many breeds of the unsympathetic environment have developed many adaptive traits that enhance survivability. Today with advancement in the scientific acumen, efforts are needed for enhancing the competence of breeding for harsh environments and also to make the future planning and execution for developing strains which can withstand the unforeseen climatic changes. Climatic stress of hot and cold extremes has helped us to develop the hardy breeds which are now resistant to the stressful environment.

Selection of Heat Tolerant/Genetic Traits for Adaptation

Selection is the only tool available with the breeders which effectively enhances the performance of the flock for the traits being selected. However, the rate of genetic improvement is slow with traditional breeding tools; there is a need to look into the genetic potential of existing species for heat tolerance. Climate change is a slow process which allows animals adapt to the prevailing environmental conditions. Breeding for climate change adaptation or mitigation will not be necessarily different from existing breeding programs; however, the problems related to measuring the phenotypes relevant to adaptation have to be overcome. Breeding indices should include traits associated with thermal tolerance, low quality feed and disease resistance, and give more consideration to genotype by environment interactions (GxE) to identify animals most adapted to specific conditions (Hoffmann, 2010).

Good environment favors high production whereas bad environment hampers it. Production of any kind, milk or meat characteristics or wool traits require congenial environment to have better genotype and environmental interactions. Selecting animals for heat tolerance needs a new understanding by livestock holders and development agencies. A fresh project at mass scale with government's efforts will add much more meaning to it.

The trend of globalization has distributed the germplasm throughout the world, which may not be suitable for future climatic changes. Developing countries which still have a high proportion of the population in rural areas are better placed for a future when localization will replace globalization as the basis of sustainable lifestyles.

Factors Accelerating Erosion of Livestock Biodiversity

Several factors are in play for the erosion of the livestock biodiversity. Focus on very few economically important breeds is one reason. World's best genetic improvement programs are employed for these few chosen breeds and many known and thousands of unknown breeds in least developed and developing countries go unsung. As a result of this, many breeds lose their potential to compete with other breeds and ultimately fall in the category of endangered population. Here the preference is given to the high-input, high-output breeds developed for benign environments of the developed countries. This works as a double-edged sword, first it criticizes the local indigenous livestock as inferior and next it dilutes the native livestock with foreign genes, least adapted for such environment. In many developing countries this activity has created the never ending series of livestock erosion problems; India is one of the best example for this phenomenon.

Conclusion

It is ascertained that the animals will try their level best to adapt with changed environments. However, the natural bred populations will have better adaptability than controlled bred populations as the former are not affected with inbreeding depression. Further, if climatic changes are intensely erratic, then the populations will undergo evolution process with modifications of physiological and anatomical characteristics. If not adapted, will go towards extinction. In the middle, the mutation process of the genes may occur to have sustainability to the animal genetic resources. But it is sure that the animals have better adaptability than human beings and we should be optimistic in our expectations that the animals will be able to serve human beings as did in the past. Favorable correlation suggests that if major importance is placed on performance traits in stressful environments, adaptability traits would not be compromised and thus the most productive and adapted animals for each environment need to be identified for breeding purposes.

References

Ames D.R., Ray D.E. (1983) Environmental manipulation to improve animal productivity. Journal of Animal Sciences. 57:209–220.

Dobzhansky T. (1970) Genetics of the evolutionary process. Columbia Universty Press, NY

Hoffmann I. (2010) Climate change and the characterization, breeding and conservation of animal genetic resources. Animal Genetics. 41(1):32–46.

Kulkarni A.A., Pingle S.S., Atakare V.G., Deshmukh AB (1998) Effect of climatic factors on milk production in crossbred cows. Indian Veterinary Journal. 75(9):846–847.

Mandal D.K., Rao A.V.M.S., Singh K., Singh S.P. (2002) Effects of macroclimatic factors on milk production in a Frieswal herd. Indian Journal of Dairy Science. 55(3):166–170.

Shinde S., Taneja V.K., Singh A. (1990) Association of climatic variables and production and reproduction traits in crossbreds. Indian Journal of Animal Science. 60(1):81–85.

Turner J.W. (1980) Genetic and biological aspects of zebu adaptability. Journal of Animal Sciences. 50(6): 1201–1205.

Wathes C.M., Jones C.D.R., Webester A.J.F. (1983) Ventilation, air hygiene and animal health. Veterinary Record. 113 (24):554–559.

20

Genetic Diversity of Livestock and Poultry under Changing Climate Scenario: Strategies and Operations

Sanat Mishra

Introduction

Diversity connotes multiplicity of variety. In a given ecosystem, diversity refers to presence of number of different species of flora and fauna that have ecological niches in agreement with the environment. Scaling up, there exists diversity between different ecosystems. All life forms of flora and fauna on earth contains genetic material, that is the repository of an immense amount of genetic information in the form of traits, characteristics, etc. Focusing on the species level, genetic diversity refers to genetic variation that is observed in a population. Thus, genetic diversity is the total number of genetic characteristics in the genetic makeup of a species. This provides in built latitude to the nature and/or the livestock breeders (including farmers) to manoeuvre/ manipulate so as evolve an improved population of desired qualities and traits. Further, genetic diversity is essential because the presence of adequate and appropriate genes in a population will make the species tide over adversaries owing to climate change and other emerging threats like new disease outbreaks. The 'desired' divergent gene pool provides the inherent resilient mechanism that equip the species to cope with likely challenging situation.

Over the years, with human interventions, the genetic diversity has been interfered with to meet specific needs; mostly with respect to garnering higher production and productivity. In the process, the diversity has been compromised for achieving the unitary goal. In certain cases, geographical barriers and other bio-physical constraints have also led to loss of diversity within a population. Studies on impact of climate change especially the effect of rise in environmental temperature on milk production of dairy animals indicate that global warming will negatively impact milk production. During 2020, the annual loss in milk production of cattle and buffaloes due to thermal stress will be around 3.2 million tonnes of milk costing more than Rs 5000 Crore at current price. Similarly, there would be prevalence of many emerging diseases including those with zoonotic implications owing to change in climatic conditions. Thus, the present day society is confronted with an imminent development dilemma of ecology vs. economics more specifically genetic diversity vs. higher production. It has led to volume of debates and discussions on the livestock development agenda to be pursued. The apparently conflicting interests has also generated antipathy amongst two schools of development.

Rationale

'Can the twain meet', the principal premise upon which this paper is based upon. This paper attempts to bridge the gulf between the two divergent 'schools of thought'. The paper is conceptualized with an overarching objective of developing strategies and operations with livestock farmer's perspective, to evolve a 'middle path' that would conjoin the beauties of both the concepts. As the Indian farmers need both; they prefer livestock species that are adaptable to less resource endowed situations, with attributes like moderately to high disease resistance, heat tolerance and simultaneously yield enough to provide gainful employment. The sustenance of genetic diversity in a given livestock population with moderate to high level of economic performance is the context of this paper. Therefore, it is intended develop the livestock species with adaptable behaviour under changing climatic conditions on one hand along with ability to produce and give optimal economic return to farmers on the other.

3. Etymology

- *Genetic material*: Any material of plant, animal, microbial or other origin containing functional units of heredity.
- *Genetic resources*: Genetic material of actual or potential economic, scientific or societal value contained within and among species. In a domesticated species, it is the sum of all the genetic combinations produced in the process of evolution.
- *Genetic distance*: A measure of the genetic differences between two populations (or species); calculated on the basis of allelic frequencies in both populations.
- *Genetic diversity*: Heritable variation within and among populations that is created, enhanced or maintained by evolutionary or selective forces.
- *Genetic variation*: Occurrence of differences among individuals of the same species, arising due to variation in alleles, genes or genotype.

- *Climatic Change:* It is any significant long-term change in the expected patterns of average weather of a region (or the whole Earth) over a significant period of time. Climate change is about abnormal variations to the climate, and the effects of these variations on other parts of the Earth. It is a change in the state of the climate that can be identified by changes in the mean (and/or the variability), and that persists for an extended period, typically decades or longer (Source: IPCC)

Situation Analysis

Population dynamics

As on date, the ICAR- National Bureau of Animal Genetic Resources (NBAGR) have registered 183 'defined breeds' of livestock and poultry in India. Based on the existing population strength, these breeds have been categorised into three groups viz. less than 5000, within 5000-10000, and more than 10000, as depicted below;

Sl. No.	Species	Defined Breeds (Nos.)	Number of Breeds with population strength		
			<5000	5000-10000	>10000
1.	Cattle	43	3	1	39
2.	Buffalo	15	3	-	12
3.	Goat	34	1	2	31
4.	Sheep	43	4	1	38
5.	Camel	09*	4		4
6.	Horse	07	4	1	2
7.	Chicken	19*	1	2	7
8.	Yak	01*			
9.	Mithun	--			
10.	Pig	08	1		7
11.	Donkey	02	1	1	
12.	Duck	01			1
13.	Geese	01*			
Total		183			

Source: www.nbagr.res.in (2019), *Data on some of the breeds not available

This analysis has revealed that population reported under certain breeds of livestock and poultry are alarmingly below 5000 numbers and some of them also range between 5000 to 10000. In Toto, 30 breeds that is more than 16 per cent of the total numbers of identified breeds are under near threatened or vulnerable categories, which urgently calls for attention. Further, for certain breeds there are data deficit, which means they have not been studied at length. There is a strong need to enumerate these breeds at an early date. More importantly, while analysing the population data, it is imperative to keep in focus the trend in rise or fall in population between two successive census. This provides insight into the overall environment (both natural and man-made) under which the specific breed is reared.

Breeding policies

The National Livestock Policy (NLP), 2013 has been formulated to provide a policy framework for improving productivity of the livestock sector in a sustainable manner, taking into account the provisions of the National Policy of Farmers, 2007 and the recommendations of the stakeholders, including the States. Under the Constitution of India, livestock development falls within the jurisdiction of the State Government. However, the Central Government supplements and complements the efforts of the State Governments through different schemes and programmes apart from creating an enabling environment to promote sustainable growth of the sector. NLP 2013 envisages that each state would prepare or revisit livestock species-specific breeding policy to ensure faster growth in production and productivity. As one date, the breeding policies adopted by different states relating to cattle and buffaloes are well documented, drafted and followed fairly well. In contrast, the same in case of small ruminants and other livestock species including poultry have ben loosely drafted or not very specific. Mostly, the breeding policies relating to small ruminants and other species do not explicitly capture the breeding requirements of these livestock and poultry. Therefore, it is imminent that each state should put in place a comprehensive species specific livestock and poultry breeding policy document that would offer scope to improve the production potential of the species on one hand and provide opportunities to the farmers to realize higher profit, on the other. Further, the state and national Governments should pay adequate attention for implementation of the breeding policy in letter and spirit. In order to make the livestock owners aware about the nuances of the policy and its implicit benefits on implementation; periodic awareness campaigns both through physical and digital space may be carried out. Further, the logistics supports viz. prescribed disease-free semen, quality AI services, proven bulls and availability of the breeding services close to livestock owners would definitely accrue long-term dividends out of the breeding programme.

Livestock Management System

It is not out of context to mention that livestock and poultry are reared in diverse agro-climatic conditions, further under multiple micro climatic situations delineated as agro-ecological situations. This underpins the fact that these animals are reared under different biophysical and socio-economic settings, wherein the resource bases vary considerably. Broadly, the multiplicity of situations can be categorized into three livestock management system (LMS) viz. intensive management, semi-intensive management and extensive management. Otherwise, based on resource endowment, these LMSs can be termed as resource endowed, resource moderate and resource deficit environments. Further, the LMS is intrinsically linked with the purpose for which these animals are reared viz. milk production, drought power, or both; meat, fibre etc. Similarly, for poultry it may be meat, egg or both. Therefore, it is imperative that the country should develop location specific strategic planning so that available resources in different agro-ecological zones of the country be exploited judiciously and utilize sustainably for further enhancing the productivity of these animals and birds.

Rashtriya Gokul Mission (RGM)

RGM is a comprehensive programme launched by Government of India during 2014-15 with an overarching objective of conserving and developing the indigenous breeds of the country in a focused and scientific manner. This mission also includes improving their genetic makeup, enhancing the milk productivity and distribution of disease free high genetic merit bulls for natural service. It aims to establish integrated indigenous cattle centres "Gokul Gram", establish Breeder's societies "Gopalan Sangh", award to farmers "Gopal Ratna" and Breeders' societies "Kamadhenu", assistance to institution which are repositories of best germplasm. The bovine genetic resource of India is represented by 43 registered indigenous breeds of cattle and 15 registered buffalo breeds. Indigenous bovine species are robust, resilient and suited to the climatic environment of their respective breeding tracts. The milk of indigenous animals is high in fat and SNF content. Awards under the mission are being bestowed since 2017-18 and so far 22 Gopal Ratna and 21 Kamdhenu awards have been distributed. Till date, 20 Gokul Grams have been sanctioned for 13 states under the scheme with an outlay of Rs 197.67 crores. Under RGM, two "National Kamdhenu Breeding Centres" (NKBC) have been established in the state of Andhra Pradesh and Madhya Pradesh; these units are functioning as 'Centres of Excellence' to develop and conserve indigenous breeds in a holistic and scientific manner. The cattle breeds included for development are Deoni, Gir, Kankrej, Red Sindhi, Sahiwal, Rathi, Tharparkar, Punganur, Ongole, Krishna Valley, Malnad Gidda, Kangeyam and that of buffalo breeds are Pandharpuri, Jaffarabadi, Banni, Mehsana and Murrah. Further, under the mission an e-market portal connecting breeders and farmers, an authentic market for quality- disease free bovine germplasm in the form of: i) semen; ii) embryos; iii) calves; iv) heifers and v) adult bovines with different agencies/stake holders have been developed. This has helped in accessing quality breeding material. Other initiatives under this mission are *Pashu Sanjivani*- an animal welfare programme with provision of animal health cards; *Advanced Reproductive Technologies*- Assisted Reproductive Technique- In-vitro Fertilization (IVF)/ Multiple Ovulation Embryo Transfer (MOET) and sex sorted semen technique to improve availability of disease free female bovines; *National Bovine Genomic Center for Indigenous Breeds (NBGC-IB)*- selection of breeding bulls of high genetic merit at a young age using highly precise gene based technology.

Environmental changes

Livestock sector is facing newer challenges, like increased incidence of emerging and re-emerging animal diseases, vulnerability to exotic diseases, shortage of feed and fodder and need to increase production to meet demand for animal products etc. Livestock species and breeds have ecological distribution and do not follow the geo-specific boundaries of the states. Therefore, sustainable livestock development requires integrated efforts across the states with an overall national perspective.

Strategies and Operations

The following strategies and operations have been delineated for maintaining genetic diversity of livestock and poultry under changing climate scenario;

- ***Registration of livestock and poultry breeds:*** Any population of livestock and poultry found in India can be registered in the form of breed, provided it fulfils the criteria laid down for the registration. A breed is defined as group of domestic livestock with definable and identifiable external characteristics that enable it to be separated by visual appraisal from other similarly defined groups within same species and/ or a group for which geographical and/or cultural separation from phenotypically similar groups has led to acceptance of its separate identity.
- ***Registration of variety / strain/ line of poultry:*** Distinctiveness characteristics of the variety/ strain/ line in comparison to other varieties/ strains/ lines available in the country need to be enlisted.
- ***Information system on AnGR of India:*** An Information System on Animal Genetic Resources of India (AGRI-IS) has been developed at ICAR-NBAGR, which covers all the indigenous breeds of domestic livestock and poultry species from India. This database contains descriptors of various breeds of livestock and poultry, information on farms, semen production, vaccine production; and district-wise information on population, animal breeding, animal health infrastructure, animal products like milk, meat, egg, wool, etc. It also stores photographs of male and female animals of breeds. Information with regards to newly registered breed is also incorporated for documentation purpose.

- ***Periodical breed-wise census of livestock population:*** Department of Agriculture and Farmers Welfare, Government of India is conducting livestock census every five-year interval with support from State Departments to understand the population dynamics and changes thereof. During last two Census, breed-wise data have been collected, which has helped to develop insight and overall strategies. However, the breed wise census can be further intensified with identification of true-to-type breeds by using clear and distinct photographs taken from all angles i.e. 360^0 coverage with focus on specific breed characteristics. This will help to record all the indigenous breed population that are true-to-type.
- ***Monitoring of indigenous breed performance:*** Along with periodical census, monitoring of the indigenous breeds by sample survey may be carried out to analyse the size of population, farmer-wise herd size, availability of breeding males per females, distribution, livestock management practices, and so on. This will help to identify the risk involved in population existence; viz. breed is not at risk, breed is vulnerable, breed is endangered, breed is critical and breed is extinct. Based on the risk status of the given population the conservation strategies will be drawn.
- ***In-situ conservation of livestock and poultry:*** Firstly, in situ conservation work requires collection of information on different aspects like establishment of institutional structure and policies including specific measures to conserve breeds at risk, population status of the breed in its native tract. The conservation programmes should go hand in hand with breed improvement programme. The genetic improvement programme should include selection of quality and disease free males for traditionally valued characteristics and adaptability to the native environment.
- ***Involvement of farmers in breed improvement programme:*** Nucleus herd should be established in respective native tract of particular breed. Farmers maintaining the animals of that breed should be given incentives for encouraging more rearing. Farmers need to be involved in the setting the criteria for selection and a comprehensive participatory breed selection plan may be conceptualized and operationalized.
- ***Establishment of indigenous breed farms:*** Exclusive farms of indigenous breeds should be declared as germplasm repositories and used for production of quality breeding males and semen. Efforts should be made to establish one such farm for each breed in the native breed tract.

- ***Unique identification of pure breeds:*** The pure breeds of indigenous origin need to identified with a unique identification number. The performance of these animals and birds need be recorded periodically for monitoring of their performances.
- ***Revision State Livestock and Poultry Breeding Policy:*** Many states have not promulgated conducive livestock breeding policies for allowing and adopting a farmer-centric, breed-centric and agro-eco situation specific breeding of livestock and poultry. In certain cases, lip-service has been provided with cosmetic changes in the archaic policy paper. The policy documents should also delineate a roadmap for implementing the strategies. Once the breeding tracts are defined in the policy document, a deterrent attempt would be adopted for discouraging the farmers to pursue crossbreeding practices for pure breeds.
- ***Promotion of indigenous breed associations:*** A special attention may be given to evolve community level breed associations for conservation and improvement of livestock and poultry breeds in their respective native tracts. Capacity building and other incentive packages may be rendered for these association to be actively involved in the decision making.
- ***Formulation of livestock and poultry breeders' right act:*** In line with PPV&FRA Act developed for protection of plant varieties and farmers' right; the livestock and poultry breeders' right should also be protected for posterity.

21

Impact of Climate Change on Poultry Health and Production

C Balachandran

Introduction

Birds lack sweat glands; hence non-evaporative cooling mechanism (radiation, conduction and convection) is the major route of heat dissipation within thermo-neutral range. Beyond this TNZ, majority of heat loss occurs as insensible heat loss *i.e.,* Panting. Panting is one of the visible responses in poultry during exposure to heat stress associated with considerable energy expenditure. This specialized form of respiration dissipates heat by evaporative cooling at the surface of the mouth and respiratory passage. Panting increase, the loss of CO_2 from the lungs, this leads to a reduction in the partial pressure of CO_2 and also reduces bicarbonate levels in blood plasma. In turn, the lowered concentration of hydrogen ions leads to rise in plasma pH, a condition generally referred to alkalosis (Daghir, 2009). Birds also respond by reducing their feed intake and thus metabolizable energy (ME) intake to reduce thermo genesis as an adoptive mechanism. If panting fails to prevent their body temperatures from rising, the birds become listless, comatose and soon die due to respiratory, circulatory or electrolyte imbalances (Saif *et al.,* 2003).

During exposure to high temperature, chicken consume less feed and more water (May and Lott, 1992). Water consumption by chicken is, however, increased during periods of heat stress to meet an increase in maintenance requirements and cool the body core temperature. The respiratory rate of birds is dependent on the age, ambient temperature and relative humidity (Syafwan *et al.,* 2011). The nasal cavity act as heat exchanger medium which helps to dissipate excess heat through evaporative cooling; hence panting is regarded as the most obvious clinical sign of heat stress in poultry.

Birds exposed to higher temperature for longer duration leads to depress the thyrotrophic axis, which lead to reduced T3 level. Heat stress leads to increased secretion of corticosterone and this increased corticosterone had a negative effect on circulating T3 level. As a response to heat stress, birds decrease the heat production in the body by decreasing the T3 levels, as T3 directly involves in the basal metabolic rate. Heat conditioned birds had a significantly improved thermo-tolerance. The improved thermo tolerance was indicated by a significantly lower metabolic rate and significantly decreased T3 levels.

Olfati *et al.* (2018) who have found that there was a significantly increased corticosterone level in the heat exposed birds when compared with the control birds. The birds subjected to high temperature, which leads to the rapid release of corticosterone (Jahejo *et al.* 2016). Corticosterone level was one of the indicators to access the stress response in the birds. Corticosterone can inhibit glucose uptake by muscle and adipose tissues, which leads to the decomposition of muscle protein and adipose tissue. This phenomenon promotes gluconeogenesis, which provides sufficient energy for the body to resist stress. This might be due to the effect of heat stress on the activation of the HPA axis.

The HSP70 considered as a highly conserved protein, will be up regulated during the stressful conditions. Thermal stress appears to increase the activity and amount of heat shock transcription factor, enhancing the HSPs mRNA synthesis leading to high concentration in heat exposed birds. The possible reason for higher expression in exposed birds might be due to the fact that, HSP genes have no intron, which hastens the transcription process after thermal stress (Rajkumar *et al.*, 2015) and it increases with gradual heat stress, resulting in higher expression of HSPs in heat stressed birds.

Heat shock proteins may have either cytoprotective or apoptotic cellular effects. The heat shock paradox suggests (DeMeester *et al.*, 2001) that if the heat shock response and HSP expression occurs prior to an inflammatory response, the subsequent effect on cells is cytoprotective. Conversely, if cells are "primed" by inflammation and then incur a heat shock response, this may lead to cell death or apoptotic cellular effect. HSPs act as activators of the innate immune response and stimulate or induce the production of pro-inflammatory cytokines, such as tumor necrosis factor-α (TNF-α), IL-1, IL-6, and IL-12 and the release of nitric oxide by monocytes, macrophages and dendritic cells (Tsan and Gao, 2004).

Production performance

Decreased body weight in heat stressed birds may be due to a reduction in feed consumption, reduced absorption and utilization of nutrients and diversion of energy resources towards stress response. When the birds were subjected to the heat stress, as a protective mechanism to reduce the internal core temperature, the feed consumption is markedly reduced to decrease the metabolic heat production, which adversely affects the body weight and weight gain. Sivaramakrishnan (2017) who have reported that there was a significantly decreased feed consumption in birds subjected to thermal stress. Increased feed intake will lead to increased metabolic heat; therefore, in order to reduce heat increment, birds might have reduced the feed intake. During heat stress, blood supply will be diverted towards the periphery for peripheral vasodilatation which will promotes heat dissipation. Hence, the blood supply to digestive tract was markedly reduced which also affects the feed consumption and leads to poor feed conversion ratio.

Thermal stressed birds often lay eggs with thin shell or shell less because of an acid-base balance disturbance as a result of panting. Birds pant to reduce their body temperature and results in respiratory alkalosis (Lin *et al.*, 2004). The higher blood pH reduces the activity of the enzyme *carbonic anhydrase*, resulting in reduced calcium and bicarbonate ions transferred from blood to the shell gland and decreasing free calcium levels in the blood. This might be a reason for the soft-shell eggs in heat stressed birds.

Mitigation Strategies to Reduce Poultry Related GHGs

Housing and manure treatment

- In open sided cage houses GHG emission will be less when compared to deep litter system.
- Liquid manure from deep-pit housing systems produces greater emissions of CH_4, N_2O and CO_2 than natural and forced dried manure from belt housing systems.
- More recently, wet scrubbers are being investigated with the aim to precipitate dust, NH_3 and odour from the exhausted air.
- ***Biodemethanation:*** *Pseudomonas methanica* produces formic acid from methanol oxidation. Vigorous oxidation of methane is noticed by *Methlyococcus capsulatus* at 50°C. This shows the potential of the organism to utilize methane. *Pseudomonas methanica* is an organism that uses methane and methanol by a pathway which is absent in facultative methanol organism. Pathway used by bacteria to utilize methane is allulose pathway.
- Frequent manure removal, also results in significantly lower NH_3 emissions from manure belt housing than manure scraper removing houses.

- ***Chemical neutralization:*** Different types of litter amendments used to mitigate ammonia pollution are as follows: acidifiers (alum, Sodium bisulphate and phosphoric acid), alkaline material ($CaCO_3$, hydrated or slaked lime $Ca(OH)_2$ or burnt lime), adsorbers (Clinoptilolite), inhibitors (Phenyl phosphorodiamidate) and microbial and enzymatic (urease enzyme) treatments.
- It is important that spreading the manure into thin layers would gives rise to higher emissions than stacking it into thicker layers.

Nutritional treatment

- Dietary manipulation can be an effective means to lower ammonia emissions by reducing excessive nitrogen excretion or change of manure pH.
- Inclusion of high-fiber ingredients in laying hen feeds lowered NH_3 emission from the manure.
- Bio alginates are successfully applied in human and veterinary medicine. Interesting thing is the ability of these bio alginates to absorb catabolic gases, particularly NH_3 , which is produced during digestion and conversion of nitrogen compounds.

TANUVAS Intervention

Thermo-tolerant genes in poultry

Genetic modification is one such approach that identifies the suitable genes and incorporates to the birds of interest. These genes may include thermo-tolerant genes, disease resistant genes and highly productive genes. Poultry industry generally looks forward for the highly productive traits to improve their profit. But as production efficiency increased the susceptibility towards heat stress and disease also increased. Eventually, several superior thermo-tolerant genes were identified by the TANUVAS researchers such as the naked neck gene (Na), frizzle gene or the (F) gene, dwarf gene or (dw) gene which made the bird resistant to heat stress through slow and reduced feathering, curling the feather so as the improve the heat dissipation and reduces body size to minimize metabolic heat production. Several other genes were also identified that increase the thermo-tolerance of birds without compromising the production potential. Moreover, heat shock proteins (HSPs) and in particular HSP70 and HSP90 are identified as ideal thermo-tolerant genes for heat stress in poultry.

Broiler chicken strains conditioned at 37°C showed significant increase in the body weight and better feed efficiency as compared to the birds that were conditioned at 39°C and control birds. Increase in body weight at market age might be due to compensatory growth occur in Nandanam broiler-3 and Naked neck cross. Sudden exposure of unconditioned birds had significantly higher corticosterone level compared to thermal conditioned birds. Among the preconditioned birds there was no significant difference observed and it indicates clearly better heat tolerance in thermal conditioned birds. At thermal challenge, the sudden exposure of unconditioned birds showed a significantly upregulated HSP-70 mRNA expression, whereas the preconditioned birds had lower HSP-70 mRNA expression compared to unconditioned birds. Among the strains, 37°C preconditioned Nandanam broiler-3 showed a significantly lower HSP-70 mRNA expression followed by Naked neck, Nandanam chicken-4 and Aseel. Whereas, at 39°C conditioned Aseel birds showed significantly lower HSP-70 mRNA expression followed by Nandanam broiler-3, Nandanam chicken-4 and Naked neck. From this result, it can be inferred that preconditioned birds able to withstand the further heat stress (Sivaramakrishnan *et al.*, 2017, 2018).

Aseel X Nandanam Chicken 4 (ANC4) and Aseel had lower respiratory rate at thermal challenge when compared to broiler strains (NN and Nandanam broiler-3 and Naked neck cross - NNB3). This might be due to the variation in the genetic makeup of the birds. Broilers are less adaptable than layer strains. Nandanam Broiler 3 of Poultry Research Station, which showed more resistance to heat. This complementary effect was noticed in Broiler 3 crosses, which reflected in Nandanam broiler-3 and Naked neck cross. The thermal conditioned birds had a significantly lower level of corticosterone than exposed group and control group birds. A significantly higher body weight was observed in broiler strains at 12th week of age (Nandanam broiler-3 and Naked neck cross) due to thermal conditioning given in early stages of life. The thermal conditioned (39°C; 2h) birds at different intervals had a better feed efficiency at 18th week of age. Synthetic cross breeds are better sustained the egg production compared to the native origins. The Aseel X Nandanam Chicken 4 (ANC4) is better placed than all other strains with respect to egg production. Similarly, cross breed broiler strain Naked neck X Nandanam broiler-3 (NNB3) had better egg production than NN. The synthetic layer strain ANC4 and broiler strain NNB3 had significantly low levels of HSPs expression in conditioned birds indicating these birds were successfully thermal conditioned with low levels of HSPs mRNA expression (Varun, 2019).

Scope for developing climate resilient poultry

With the applications of advanced biotechnological tools such as Functional genomics, microarray technology, Single Nucleotide Polymorphisms (SNPs), whole genome association studies, whole genome Transcriptome studies and next generation sequencing, it is possible to identify the various transcripts associated with stress pathways. These technologies offers huge scope to identify various transcripts associated with productive, adaptive and disease resistance. Using these technologies, it is possible to identify few thousands of genes which impart thermo-tolerance to poultry. Such transcripts can be validated using real time PCR and may be narrowed down to the handful numbers of reliable biological markers which can be incorporated in breeding programs and evolve a new thermo-tolerant breed using marker assisted selection.

Whole genomic Analysis for TANUVAS Aseel and Nandanam Chicken 4 (NC4) is under progress to study more than 50 economically important genes related to production and climate resilience like heat and disease tolerance.

References

Alexandratos, N., Bruinsma, J., 2012. World agriculture towards 2030/2050: the 2012 revision. ESA Working paper No. 12–03. FAO, Rome.

Aydinalp, C., Cresser, M.S., 2008. The effects of climate change on agriculture. Agric. Environ. Sci. 5, 672–676.

Barati, F., Agung, B., Wongsrikeao, P., Taniguchi, M., Nagai, T., Otoi, T., 2008. Meiotic competence and DNA damage of porcine oocytes exposed to an elevated temperature. Theriogenology 69, 767–772.

Bellarby, J., Tirado, R., Leip, A., Weiss, F., Lesschen, J.P., Smith, P., 2013. Livestock greenhouse gas emissions and mitigation potential in Europe. Glob. Change Biol. 19, 3–18.

Berman, A.J., 2005. Estimates of heat stress relief needs for Holstein dairy cows. J. Anim. Sci. 83, 1377–1384.

Bernabucci, U., Lacetera, N., Ronchi, B., Nardone, A., 2002. Markers of oxidative status in plasma and erythrocytes of transition dairy cows during hot season. J. Dairy Sci. 85, 2173–2179.

Bernabucci, U., Lacetera, N., Basirico, L., Ronchi, B., Morera, P., Seren, E., Nardone, A., 2006. Hot season and BCS affect leptin secretion of periparturient dairy cows. Dairy Sci. 89, 348–349.

Bhatta. R, V. Sejian and P. M. Malik, 2016. Livestock and climatic change: Contribution, impact and adaptation from India contest, Innovative designs, implements for global environment and entrepreneurial needs optimizing utilitarian sources, INDIGENOUS, 94-110.

Chase, L.E., 2012. Climate change impacts on dairy cattle. Climate change and agriculture: Promoting practical and profitable responses.

Chhabra, A., Manjunath, K.R., Panigrahy, S., Parihar, J.S., 2013. Greenhouse gas emissions from Indian livestock. Climatic Change 117, 329–344.

Daghir, N. J., 2009. Nutritional Strategies to Reduce Heat stress in Broiler and Broiler Breeders. Lohaman Information. 44(1): 6.

DeMeester, S. L., T. G. Buchman and J. P. Cobb, 2001. The heat shock paradox: does NF-kappaB determine cell fate? FASEB J. 15(1): 270 - 274.

Downing, M.M.R., A. Pouyan Nejadhashemi, Timothy Harrigan, Sean A. Woznicki, 2017. Climate change and livestock: Impacts, adaptation, and mitigation, Climate Risk Management 16: 145–163.

FAO (Food and Agriculture Organization of the United Nations), 2009a. Global agriculture towards 2050. High Level Expert Forum Issues Paper. FAO, Rome.

Fathi, M. M., A. Galal, S. El-safty and M. Mahrous,2013. Naked neck and frizzle genes for improving chickens raised under high ambient temperature: I. Growth performance and egg production, World's Poultry Science Journal , 69: 813-831.

Finocchiaro, R., van Kaam, J., Portolano, B., Misztal, I., 2005. Effect of heat stress on production of dairy sheep. J. Dairy Sci. 88, 1855–1864.

Gerber, P.J., Steinfeld, H., Henderson, B., Mottet, A., Opio, C., Dijkman, J., Falcucci, A., Tempio, G., 2013. Tackling Climate Change Through Livestock: A Global Assessment of Emissions and Mitigation Opportunities. FAO, Rome.

Haun, G.L., 1997. Dynamic responses of cattle to thermal heat loads. J. Anim. Sci. 77, 10–20.

Henry, B., Charmley, E., Eckard, R., Gaughan, J.B., Hegarty, R., 2012. Livestock production in a changing climate: adaptation and mitigation research in Australia. Crop Pasture Sci. 63, 191–202.

Hurst, P., Termine, P., Karl, M., 2005. Agricultural workers and their contribution to sustainable agriculture and rural development. FAO, Rome.

IPCC (Intergovermental Panel on Climate Change), 2007. Climate Change 2007: Synthesis Report. In: Pachauri, R.K., Reisinger, A. (Eds.), Contribution of Working Groups I, II and III to the Fourth assessment report of the Intergovernmental Panel on Climate Change. IPCC, Geneva, Switzerland, p. 104.

IPCC (Intergovermental Panel on Climate Change), 2014. Climate Change 2014: impacts, adaptation, and vulnerability. part A: global and sectoral aspects. In: Field, C.B., Barros, V.R., Dokken, D.J., Mach, K.J., Mastrandrea, M.D., Bilir, T.E., Chatterjee, M., Ebi, K.L., Estrada, Y.O., Genova, R.C., Girma, B., Kissel, E.S., Levy, A.N., MacCracken, S., Mastrandrea, P.R., White, L.L. (Eds.), Contribution of Working Group II to the Fifth Assessment Report of the Intergovernmental Panel on Climate Change. Cambridge University Press, Cambridge, United Kingdom and New York, NY, USA, p. 1132.

Jahejo, A. R., N. Rajput, N. M. Rajput, I. H. Leghari, R. R. Kaleri, R. A. Mangi, M. K. Sheikh and M. Z. Pirzado, 2016. Effects of Heat Stress on the Performance of Hubbard Broiler Chicken. Cells, Animal and Therapeutics, 2(1): 1 - 5.

King, J.M., Parsons, D.J., Turnpenny, J.R., Nyangaga, J., Bakari, P., Wathes, C.M., 2006. Modellig energy metabolism of Friesians in Kenya smallholdings shows how heat stress and energy deficit constrain milk yield and cow replacement rate. Anim. Sci. 82, 705–716.

Kunavongkrita, A., Suriyasomboonb, A., Lundeheimc, N., Learda, T.W., Einarsson, S., 2005. Management and sperm production of boars under differing environmental conditions. Theriogenology 63, 657–667.

Lacetera, N., Bernabucci, U., Ronchi, B., Nardone, A., 2003. Physiological and productive consequences of heat stress: The case of dairy ruminants. Proc. of the Symposium on Interaction between Climate and Animal Production: EAAP Technical Serie, pp. 7, 45–60.

Lin, H., K. Mertens, B. Kemps, T. Govaerts, B. De Ketelaere, J. De Baerdemaeker, E. Decuypere1 and J. Buysel, 2004. New approach of testing the effect of heat stress on eggshell quality: mechanical and material properties of eggshell and membrane. Br Poult Sci. 45(4): 476 - 482.

Mader, T.L., 2003. Environmental stress in confined beef cattle. J. Anim. Sci. 81, 110–119.

May, J. D and B. D. Lott, 1992. Feed and water consumption patterns of broilers at high environmental temperature. Poult Sci. 71: 331-336.

McDowell, R.E., 1968. Climate versus man and his animals. Nature 218, 641–645.

Mosier, A., Kroeze, C., Nevison, C., Oenema, O., Seitzinger, S., Van Cleemput, O., 1998. Closing the global N2O budget: Nitrous oxide emissions through the agricultural nitrogen cycle: OEDC/IPCC/IEA phase II development of IPCC guideline for national greenhouse gas methodology. Nutr. Cycl. Agroecosyst. 52, 225–248.

Mosier, A., Wassmann, R., Verchot, L., King, J., Palm, C., 2004. Methane and nitrogen oxide fluxes in tropical agricultural soils: sources, sinks and mechanisms. Environ. Dev. Sustainability 6, 11–49.

Nardone, A., Ronchi, B., Lacetera, N., Ranieri, M.S., Bernabucci, U., 2010. Effects of climate change on animal production and sustainability of livestock systems. Livest. Sci. 130, 57–69.

Olfati Ali, Ali Mojtahedin, Tayebeh Sadeghi, Mohsen Akbari and Felipe Martínez-Pastor, 2018. Comparison of growth performance and immune responses of broiler chicks reared under heat stress, cold stress and thermoneutral conditions. Spanish J Agri Res. 16(2): 505 - 512.

O'Mara, F.P., 2012. The role of grasslands in food security and climate change. Ann. Bot-London 110, 1263–1270.

Rajkumar, U., A. Vinoth, M. Shanmugam, K. S. Rajaravindra, and S. V. Rama Rao, 2015. Effect of embryonic thermal exposure on heat shock proteins (Hsps) gene expression and serum T3 concentration in two broiler populations. Anim. Biotechnol. 26(4): 260 - 267.

Randolph, S.E., 2008. Dynamics of tick-borne disease systems: minor role of recent climate change. Rev. Sci. Technol. Oie. 27, 367–381.

Ronchi, B., Bernabucci, U., Lacetera, N., Verini Supplizi, A., Nardone, A., 1999. Distinct and common effects of heat stress and restricted feeding on metabolic status in Holstein heifers. Zootec. Nutr. Anim. 25, 71–80.

Rosegrant, M.W., Fernandez, M., Sinha, A., 2009. Looking into the future for agriculture and AKST. In: McIntyre, B.D., Herren, H.R., Wakhungu, J., Watson, R.T. (Eds.), International Assessment of Agricultural Knowledge, Science and Technology for Development (IAASTD). Agriculture at a crossroads, Island Press, Washington, DC, pp. 307–376.

Saif, Y. M., H. J. Barnes, J. R. Gilsson, A. M. Fadly, L. R. McDonald and D. E. Swayne, 2003. In: Disease of poultry. 11th Ed., lowa State Press, A Blackwell Publishing Co. Ames, lowa. p. 1056.

Sejian, V., J. B. Gaughan, R. Bhatta and S.M.K. Naqvi, 2016. Impact on climate change on livestock productivity. www.feedipedia.org.

Sivaramakrishnan. S., A.V. Omprakash, S. Ezhilvalavan, K.G. Tirumurugaan and A. Varun, 2017. Successful development of thermo tolerance in Nandnam Broiler-3 chicken through thermal conditioning, Proceedings of Indian Poultry Science Association Conference 28-30th November 2017, pp- 95.

Sivaramakrishnan. S., A.V. Omprakash, S. Ezhilvalavan, K.G. Tirumurugaan and A. Varun, 2017. Effect of thermal conditioning on the production performance of different strains of chicken, Proceedings of Indian Poultry Science Association Conference 28-30th November 2017, pp-227.

Sivaramakrishnan. S., A.V. Omprakash, S. Ezhil Valavan, K.G.Thirumurugan and A.Varun, 2018. Effect of Thermal Conditioning on the Production Performance of different Strains of Chicken, International Journal of Livestock Research, 8 (2): 291-298.

Sivaramakrishnan. S., A.V. Omprakash, S. Ezhilvalavan, K.G. Tirumurugaan, G. Srinivasan and A. Varun, 2018. Evaluation of Heat Tolerance Potential by the Expression of HSP-70 mRNA in Different Chicken Strains through Thermal Conditioning, National conference on Native chicken production: opportunities for conservation, productivity enhancement and commercial exploitation in view of global warming held at Madraas Veterinary College, Chennai on 19th & 20th December 2018.

Sivaramakrishnan. S., A.V. Omprakash, S. Ezhilvalavan, K.G. Tirumurugaan, G. Srinivasan and A. Varun, 2018. Evaluation of triiodothyronine and corticosterone hormone in early thermal conditioned chicken strains under the thermal conditioning and thermal challenge, National conference on Native chicken production: opportunities for conservation, productivity enhancement and commercial exploitation in view of global warming held at Madraas Veterinary College, Chennai on 19th & 20th December 2018.

Steinfeld, H., Gerber, P., Wassenaar, T., Castel, V., Rosales, M., Haan, C., 2006. Livestock's Long Shadow: Environmental Issues and Options. FAO, Rome.

Syafwan, S., R. P. Kwakkel and M. W. A. Verstegen, 2011. Heat stress and feeding strategies in meat-type chickens. Worlds Poult. Sci. J. 67: 653 - 673.

Thomas, C.D., Cameron, A., Green, R.E., Bakkenes, M., Beaumont, L.J., Collingham, Y.C., Erasmus, B.F.N., de Siqueira, M.F., Grainger, A., Hannah, L., Hughes, L., Huntley, B., van Jaarsveld, A.S., Midgley, G.F., Miles, L., Ortega-Huerta, M.A., Peterson, A.T., Phillips, O.L., Williams, S.E., 2004. Extinction risk from climate change. Nature 427, 145–148.

Thornton, P.K., Van de Steeg, J., Notenbaert, A., Herrrero, M., 2009. The impacts of climate change on livestock and livestock systems in developing countries: A review of what we know and what we need to know. Agric. Syst. 101, 113–127.

UN (United Nations), 2013. World population projected to reach 9.6 billion by 2050. United Nations Department of Economic and Social Affairs.

Varun, A., 2019. Influence of heat shock proteins on the production and immunity parameters in native chicken germplasm, Ph.D., thesis submitted to the TANUVAS, Chennai (unpublished data).

Varun, A., A.V. Omprakash, K. Kumanan, S. Vairamuthu and N. Karthikeyan. Effect of thermal conditioning on physiological response of native chicken. The Indian Veterinary Journal, June 19, Accepted.

Varun, A., C. Yogesh, P.V. Sangeetha, S. Ezhil Valavan, M. Thangapandiyan, C. Pandian and A.V. Omprakash. Diagnosis of lymphoid leucosis in an organized aseel farm. The Indian Veterinary Journal, May 19, Accepted.

Webster, A.J., 1991. Metabolic responses of farm animals to high temperature. In: Ronchi, B., Nardone, A., Boyazoglu, J. (Eds.), Animal Husbandry in Warm Climates. EAAP Publication, pp. 15–22.

22

Effect of Climate Change on Sustainable Backyard Poultry Production

N C Behura and L Samal

Introduction

The Intergovernmental Panel on Climate Change Fifth Assessment Report (IPCC, 2013) emphasized that warming of the climatic system is unequivocal and that anthropogenic warming will continue for centuries to come due to timescales associated with climate processes and feedbacks. While stresses caused by climate change are likely to negatively impact the agricultural and livestock sector and escalate existing small-scale farming challenges; agricultural and livestock productivity and sustainability have to improve to meet the needs of the growing human population (DEA, 2013; Zamykal and Everingham, 2009). Climate change is seen as a major threat to the survival of many species, ecosystems and the sustainability of poultry production systems in many parts of the world. Nevertheless, global demand for poultry products is expected to increase, as a result of the growing human population, and its growing affluence. The extreme climatic conditions impose various stresses on poultry birds which adversely affect their growth, production and reproduction.

While climate change is a global phenomenon, its negative impacts are more severely felt by poor people in developing countries who rely mostly on the natural resource base for their livelihoods. Moreover, rural poor communities rely greatly for their survival on agriculture and poultry which belong to the most climate-sensitive economic sectors. Climate change is expected to intensify existing problems and create new combinations of risks and shift hazard zones. The incidences of draughts, snow-storms and blizzard like events have risen. The situation is made worst due to factors such as widespread poverty, inequitable land distribution, limited access to capital and technology inadequate infrastructure, long term weather forecasts and inadequate research and extension. Intensive poultry production has much less impact on global warming than organic or free-range production (Van der Sluis, 2007). Seo and Mendelsohn (2006) reported that small farms of poultry are better able to adapt to warming. However, organic egg production needs more energy than non-organic and increases most environmental burdens. The most disadvantageous, from environmental point of view, is litter-free breeding of birds which causes great amounts of liquid manure.

Climate change

Global temperatures have substantially increased over recent years, such that between 1980 and 2016, increases of over 1°C have been recorded over vast terrestrial regions. Extreme weather events have increased and regional climate patterns are changing. In the twentieth century the average temperature increased by 0.74°C, sea level increased by 17 cm and a large part of the Northern Hemisphere snow cover vanished (IPCC, 2007). The IPCC (2001) in its global climate scenarios has reported that the future pattern of global warming is going to be more evident at high altitude zones. Especially the tropics and sub-tropics of which the Himalayas is the largest range where there will be faster warming of 3 to 5 times more than the rest of the world. Alpine glaciers in the Himalayas are found to be very sensitive indicators of climate change. An observation by the Working Group on Himalayan Glaciology of the International Commission for Snow and Ice (ICSI) in 1999 stated that glaciers in the Himalayas are receding faster than in any other parts of the world and if the present rate continues, the likelihood of them disappearing by the year 2035 is very high. Investigations have shown an overall reduction in glacier area from 2077 sq. km in 1962 to 1628 sq. km in the year 2007, an overall deglaciation of 21% while the number of glaciers has increased due to fragmentation (IPCC, 2007). This is going to create a catastrophic situation for India. Such temperature increases, even at the lower end of the range, are expected to have far-reaching impacts on food production and food security. Small-scale and subsistence farming in developing countries is expected to suffer complex and localized consequences due to climate change and constrained adaptive capacity (Archer van Garderen, 2011; Shisanya and Mafongoya, 2016).

Importance of backyard poultry

Poultry production and consumption has been increased in the world. Poultry meat accounts for about 33% (87% chicken and 6.7% Turkey) of the global meat consumption (FAO, 2010). The worldwide average per capita consumption of poultry is around 11 kg and poultry meat and eggs are highly nutritious. The indigenous poultry population of Odisha is 57,17,851 (19th Livestock Census, 2012). Odisha stands 10th in terms of indigenous poultry for households and number of households maintaining indigenous poultry after North-Eastern states, West Bengal and Jammu and Kashmir. Fourteen (14) out of 30 districts are tribal dominated and backyard poultry rearing is common practice influencing their socio-cultural life. Backyard poultry also plays significant role for adding sustainability to the livelihood of poor farmers.

Poultry production is a predominant global farming sub-sector, and has been an important food source for humans for over 8000 years. Poultry are birds that include fowl, turkey, duck, goose, ostrich, guinea fowl, etc which render not only economic services but contribute significantly to human food as a primary supplier of meat, egg, raw materials to industries (feathers, waste products), source of income and employment to people compared to other domestic animals (Demeke, 2004). Poultry production is common in rural areas and mainly constitutes nondescript village chicken breeds. These chickens are also commonly referred to as 'rural', 'backyard', 'indigenous', 'scavenging', 'traditional', 'local', 'native' or 'family' chickens. Poultry plays a significant role in the livelihoods of rural communities, supplementing quality food in the form of meat and eggs, as well as secondary income. Although rural poultry production is not an occupation per se, it often supplements household income, primarily for poor families. According to Reta (2009), chickens are also considered as the last resource for poor farmers, indicating that these are often the last capital items that households have left when transitioning into a state of poverty. Poultry further has socio-cultural functions in everyday village life, such as providing chickens as gifts or as tokens of appreciation for services rendered (Muchadeyi *et al*., 2004). Poultry are efficient converters of feed to egg and meat within a short period of time. In terms of nutritive value, poultry egg rank second to cow milk. Poultry in rural areas contributes towards the mitigation of HIV/ AIDS through improved food security and income generation. In addition, because these birds are hardy, farmers recognize considerable production benefits with minimum input in terms of time commitment, feed, housing and general health management aspects. Several researchers have argued that strengthening rural poultry production may help overcome poverty and malnutrition in impoverished communities. However, poultry production in many developing communities is frequently characterized by management challenges and low productivity, and is faced with several other constraints. Of growing concern is that climate change may be impacting on poultry growth and development, and ultimately overall poultry production. Of all bioclimatic parameters, temperature is considered the most important factor affecting poultry birds, as it has a direct impact on them often causing heat stress.

Climate change challenges faced by backyard poultry

Since the 1980s, there has been considerable interest in the relationship between commercial poultry production and heat stress, from establishing the impacts of heat stress to intervention strategies. Some earlier studies have associated selection for rapid growth with increased susceptibility of poultry (particularly broilers) to heat stress, given that with rapid growth rates, the basal metabolic rate of broilers is increased. In turn, this generates more heat, which may eventually cause bird mortality. Since rural poultry is characterized by slow growth rates, it is possible that they are less susceptible to the impact of heat stress. However, recent research suggests that all birds generally react in a similar manner to heat stress, but may express individual variation of intensity and duration of responses (Lara and Rostagno, 2013). This concurs with earlier studies (Boissy *et al.*, 2007; Hemsworth, 2003) arguing that variations might occur when other stressors such as limited housing space and insufficient ventilation accompany heat stress.

Rural backyard poultry production faces a variety of challenges, including amongst others such as inherent slow growth rates, poor egg production, poor nutrition, lack of adequate housing and health care, predation and high mortalities. Such challenges pose a significant threat to the sustainability of rural poultry production in developing regions. Backyard poultry mainly depends on scavenging for nutritional needs, although supplementary feed is sometimes provided. Supplementary feeding is usually irregular, non-selective and characterized by birds at different stages of growth competing for the same feed. This results in weaker groups, such as chicks, receiving inadequate nutrition. In addition to this backyard poultry system affecting nutrition, it may further expose birds to parasitism and predation during scavenging (Nyoni and Masika, 2012). The important environmental stresses include *heat stress* due to direct effects of high temperature, relative humidity and solar radiation and *cold stress* at high altitude regions both resulting in *nutritional stress* due to the negative effects on pasture growth, reducing quantity and quality of feed and *disease stress* due to spreading of harmful microorganisms into new regions. The climate change is becoming increasingly evident across most regions globally, and may have adverse effects on the welfare of poultry. Observed increases in temperatures are likely to cause heat stress on poultry, especially in rural condition where there is limited understanding on how environmental conditions impact poultry, and consequently on the associated best management practices.

Heat stress

Several biotic factors such as a species' genetic potential, life stage and nutritional status, may determine the level of vulnerability to heat stress. According to Huey *et al.* (2012), species with high physiological capacities to buffer environmental variations and high potential for rapid evolution are better equipped to survive rapid warming. The level of tolerance to high environmental temperatures is also determined by the species' place of origin. Poultry birds that have not been exposed to high temperatures are generally more vulnerable to heat stress than those that have adapted (acclimatized) over time. Heat stress is one of the most important environmental challenges adversely affecting poultry production worldwide. Understanding and buffering environmental conditions is thus imperative to successful poultry production.

Both of the climate change and poultry productions have always negative impacts one over the other. Poultry flocks are particularly vulnerable to heat stress because birds can only tolerate narrow temperature ranges. Heat stress could affect poultry productions due to the impacts of increasing air temperature, feed-grain availability and favoring the diseases. The challenges posed by climate change are broadly divided into two categories: loss of productivity or increasing costs. Regarding productivity, housing systems need to be managed to maintain optimal seasonal temperatures and reduce the risk of heat stress, and increased investment will be required in ventilation and cooling. Reproductive capacity may decrease. It is possible that the observed low poultry output (in terms of eggs produced and number of birds consumed and/or sold) in some small-scale rural poultry farming regions may, in part, be a function of heat stress.

In commercial production systems where environmental microclimates are well controlled, local climate (or climate change) may not be of significant concern. However, in rural areas of developing countries, it is likely that rural poultry is adversely affected (i.e. stressed) by extreme environmental conditions, as birds continue to interact with the local environment during scavenging, and farmers have less capacity to control their living environments. Winsemius *et al.* (2014) reported that subsistence farming in developing countries is vulnerable to extreme weather conditions and that livestock in rural areas may be critically impacted during exceptionally hot days, consequently impeding productivity.

Stressful heat conditions not only reduce feed intake and body weight, but also affect reproductive performance of laying hens by interrupting egg production (Abidin and Khatoon, 2013). Earlier studies ascertained that heat stress disrupts hormones responsible for ovulation and decreases responsiveness of granulosa cells to luteinizing hormone in hens (Novero *et al.*, 1991). In addition, egg production, egg weight, shell weight and shell thickness are considerably compromised by heat exposure, resulting in poor egg quality (Renaudeau *et al.*, 2012). Given standard room temperatures of between 20 and 24°C, normal respiration rates of adult birds will range between 20 to 59 breaths per minute (Marchini *et al.*, 2007). However, as ambient temperature increases, birds tend to increase their respiration rate to as much as 151 breaths per minute for heat stressed broilers (Nascimento *et al.*, 2012). Although panting helps birds lose excess body heat to the environment (Lara and Rostagno, 2013), studies on breeders and layers demonstrated that increased panting ultimately reduces blood bicarbonate availability for eggshell mineralization, consequently impacting negatively on egg production (Mashaly *et al.*, 2004; Renaudeau *et al.*, 2012).

Disease stress

Guis *et al.* (2011) reported that climate change will alter global disease distribution. Elijah and Adedapo (2006) reported that climate change has an effect with poultry feed intake, encourage outbreak of poultry diseases which invariably reduce egg production. Gilbert *et al.* (2008) reported that little is known about the direct effect of climate change factors on highly pathogenic avian influenza transmission of domestic birds and persistence to allow inference about the possible effects.

Nutritional stress

Due to climate change and population growth, food-feed competition is an event and that leads to another concern of poultry production. According to the Rosenzweig *et al.* (1993) reports, the effect of climate change on crop yields is more adverse. Chadd (2008) reported also that additional legislation will affect most aspects of the feed sector, including those pertaining to environmental protection, feed hygiene and those linked to food-safety issues throughout the poultry supply chain.

Adaptation strategies against climate change

As with other livestock, birds maintain a relatively constant internal temperature through physiological and behavioural thermoregulation. Physiological thermoregulation entails changes in metabolism to control body temperature. For example, when ambient temperature reaches a critical upper temperature, birds may reduce voluntary feed intake to maintain a constant body temperature. In contrast, behavioural thermoregulation involves changes in posture, orientation and/or environment in order to retain a constant body temperature. Ambient temperatures above 30°C generally impact negatively on chickens and cause a reduction in feed intake and body weight, and on occasion, high mortality of broilers. In addition, major attributes of poultry meat quality (i.e. appearance, texture, juiciness, flavour, and functionality) are significantly decreased in heat-stressed broilers.

Although rural poultry is generally known to be hardy and well adapted to harsh, stressful environments, scientific information that backs these claims is limited. Since rural poultry is usually exposed to prolonged high temperatures (in hot regions) during scavenging, it is likely that they have genetically adapted to a warming climate. One behavioural strategy for coping with stressful heat conditions could be seeking shade, drinking water and changing posture to maintain core body temperature under hot environmental conditions. Although this significantly reduces radiant heat load, actively seeking shade during the day may imply inadequate feed intake, especially for backyard poultry which depend heavily on scavenging for their nutritional needs. It is likely that as rural poultry dissipate excess heat, they may pant excessively, especially during summer when highest ambient temperatures are experienced. Costs are likely to increase as the result of the need to cool buildings more in summer and reduce house humidity. Building infrastructure and maintenance will have to cope with more intense weather events and increased rainfall. This means that building plans need to consider more sustainable options, with greater investment in drainage systems to accommodate more extreme and frequent floods and frequent rainfall. Stocking density in the house may need to be reduced in extreme temperatures, and actively controlled ventilation could become essential in transportation vehicles. Winter energy costs may reduce as warmer winters reduce the need to heat buildings and flocks can be acclimatized outside. Poultry farmers should reconsider building design in new builds to more effectively cope with new climate and weather extremes, including the installation of more/new equipment to cope with new climate extremes. These include the installation of renewable energy such as solar or wind power to power poultry sheds, and using biomass boilers or anaerobic digestion of poultry litter.

Susceptibility of birds to heat stress is not determined by genetic differences in body size and age, but rather by domestication and selective breeding (Soleimani *et al.*, 2011). In addition, Soleimani *et al.* (2011) explain how selection for fast growth correlates with vulnerability to heat stress. Other studies (Felver-Gant *et al.*, 2012), however, have suggested that much variation is genetically based, and that genetic selection could be a useful strategy for reducing the impact of stress on birds.

Conclusions and future strategies

Climate change is rapidly emerging as a global critical development issue affecting many sectors in the world and is considered to be one of the most serious threats for sustainable poultry production. Poultry population is already under pressure from climate stresses which increase vulnerability to further climate change and reduce their adaptive capacity. Developing skill, resources and infrastructure for poultry industry should be aimed to understand the impacts of climate change on their growth, production and reproduction. Furthermore, improving and strengthening human capital through education, outreach and extension services, improves decision-making capacity at every level and increases the collective capacity to adapt against climate change. Efficient and affordable adaptation practices need to be developed for the rural poor who are unable to afford expensive adaptation technologies.

References

Abidin, Z., & Khatoon, A. (2013). Heat stress in poultry and the beneficial effects of ascorbic acid (vitamin C) supplementation during periods of heat stress. World's Poultry Science Journal, 69(1), 135–152.

Archer van Garderen, E. R. M. (2011). Reconsidering cattle farming in Southern Africa under a changing climate. Weather, Climate, and Society, 3(4), 249–253.

Boissy, A., Manteuffel, G., Jensen, M. B., Moe, R. O., Spruijt, B., Keeling, L. J., Aubert, A. (2007). Assessment of positive emotions in animals to improve their welfare. Physiology and Behavior Journal, 92, 375–397.

Chadd, S.(2008). Future trends and developments in poultry nutrition. Proceedings of the International Conference of Poultry in the 21st Century Avian Influenza and Beyond, November 5-7, 2007, FAO, Bangkok.

DEA. (2013). Department of Environmental Affairs: Long-Term Adaptation Scenarios Flagship Research Programme (LTAS) for South Africa. Climate Trends and Scenarios for South Africa. Pretoria, South Africa.

Demeke, S. (2004). Egg production and performance of local white leghorn hens under intensive and rural household conditions in Ethiopia. Livestock Research for Rural Development, 16(2).

Elijah, O.A. and A. Adedapo (2006). The effect of climate on poultry productivity in Ilorin Kwara state, Nigeria. Int. J. Poult. Sci., 5: 1061-1068.

FAO. (2010). Poultry Meat and Eggs: Agribusiness Handbook. Director of Investment Centre Division, FAO., Rome, Italy, Pages: 77.

Felver–Gant, J.N., Mack, L.A., Dennis, R.L., Eicher, S.D., Cheng, H.W. (2012). Genetic variations alter physiological responses following heat stress in 2 strains of laying hens. Poult. Sci. 91, 1542–1551

Gilbert, M., J. Slingenbergh and X. Xiao (2008). Climate change and avian influenza. Rev. Sci. Tech., 27: 459-466.

Guis, H., C. Caminade, C. Calvete, A.P. Morse, A. Tran and M. Baylis, (2011). Modelling the effects of past and future climate on the risk of bluetongue emergence in Europe. J. R. Soc. Interface, , 10.1098/rsif.2011.0255

Hemsworth, P. H. (2003). Human-animal interactions in livestock production. Applied Animal Behaviour Science, 81, 185–198.

Huey, R. B., Carlson, M., Crozier, L., Frazier, M., Hamilton, H., Harley, H., Honag, A. & Kingsolver, J. G. (2002) Plants versus animals: do they deal with stress in different ways? Integr. Comp. Biol. 42, 415–423.

Intergovernmental Panel on Climate Change (IPCC). 2001. Climate Change 2001: Impacts, Adaptation & Vulnerability: Contribution of Working Group II to the Third Assessment Report of the IPCC. In: J.J. McCarthy, O.F. Canziani, N.A. Leary, D.J. Dokken and K.S. White, eds. Cambridge, UK: Cambridge University Press. 1000 pp.

Intergovernmental Panel on Climate Change (IPCC). 2007. Climate Change 2007: Impacts, Adaptation and Vulnerability. Contribution of Working Group II to the Fourth Assessment Report of the IPCC. In: M.L. Parry, O.F. Canziani, J.P. Palutikof, P.J. van der Linden and C.E. Hanson, eds. Cambridge University Press, Cambridge, UK, 976 pp.

International Commission for Snow and Ice (ICSI). 1999. A report by the Working Group on Himalayan Glaciology (WGHG) of the International Commission for Snow and Ice (ICSI).

IPCC (Intergovermental Panel on Climate Change) Climate Change 2013: The physical science basis. F. Stocker, D. Qin, G.K. Plattner, M. Tignor, S.K. Allen, J. Boschung, A. Nauels, Y. Xia, V. Bex, P.M. Midgley (Eds.), Contribution of Working Group I to the Fifth Assessment Report of the Intergovernmental Panel on Climate Change, Cambridge University Press, Cambridge, United Kingdom and New York, NY, USA (2013), p. 1535.

Lara L.J., Rostagno M.H. (2013). Impact of heat stress on poultry production. Animals, 3: 356–369.

Marchini, C. F. P., Silva, P. L., & Birth, M. R. B. M. (2007). Cloacal respiration and temperature in broilers subjected to high cyclic environment temperature. Archives of Veterinary Science, 12(1), 41–46

Mashaly, M. M., Hendricks 3rd, G. L., Kalama, M. A., Gehad, A. E., Abbas, A. O., & Patterson, P. H. (2004). Effect of heat stress on production parameters and immune responses of commercial laying hens. Poultry Science, 83(6), 889–894.

Muchadeyi, F. C., S. Sibanda, N. T. Kusina, J. F. Kusina and S. Makuza. (2004). The village chicken production system in Rushinga District of Zimbabwe. Livestock Res. Rural Dev. 16 (40).

Nascimento, S. T., Silva, I. J. O., Mourão, G. B., and Castro, A. C. D. (2012). Bands of respiratory rate and cloacal temperature for different broiler chicken strains. Revista Brasileira de Zootecnia, 41(5), 1318–1324.

Novero, R. P., Beck, M. M., Gleaves, E. W., Johnson, A. L., & Deshazer, J. A. (1991). Plasma progesterone, luteinizing hormone concentrations and granulose cell responsiveness in heat-stressed hens. Poultry Science, 70, 2335–2339.

Nyoni, N. M. B., & Masika, P. J. (2012). Village chicken production practices in the Amatola Basin of the Eastern Cape Province, South Africa. African Journal of Agricultural Research, 7(17), 2647–2652

Renaudeau, D., Collin, A., Yahav, S., de Basilio, V., Gourdine, J. L., and Collier, R. J. (2012). Adaptation to hot climate and strategies to alleviate heat stress in livestock production. Animal, 6 (5), 707–728.

Reta, D. (2009) Understanding the role of indigenous chickens during the long walk to food security in Ethiopia. Livest Res Rural Dev 21: 116.

Rosenzweig, C., M.L. Parry, G. Fischer and K. Frohberg, (1993). Climate Change and World Food Supply. Oxford University, Oxford Environmental Change Unit, UK.

Seo, S.N. and Mendelsohn, R., (2006). Climate change impacts on animal husbandry in Africa: A Ricardian analysis. CEEPA Discussion Paper No. 9, Centre for Environmental Economics and Policy in Africa, University of Pretoria, South Africa.

Shisanya, S., and Mafongoya, P. (2016). Adaptation to climate change and the impacts on household food security among rural farmers in uMzinyathi District of Kwazulu-Natal, South Africa. Food Security, 8(3), 597–608.

Soleimani, A.F., Zulkifli, I., Omar, A.R., Raha, A.R. (2011). Physiological responses of 3 chicken breeds to acute heat stress. Poult. Sci. 90, 1435–1440.

Van der Sluis, W. (2007). Intensive poultry production better for global warming. Wrld Poult. 23 (2), 28-30.

Winsemius, H.C., Dutra, E., Engelbrecht, F.A., Archer van Garderen, E., Wetterhall, F., Pappenberger, F., and Werner, M. G. F. (2014). The potential value of seasonal forecasts in a changing climate. Hydrology and earth system sciences discussions. European Geosciences Union, 10, 14747–14782.

Zamykal, D., and Everingham, Y. L. (2009). Sugarcane and precision agriculture: Quantifying variability is only half the story- A review. Climate Change, Intercropping, Pest Control and Beneficial, Microorganisms, Sustainable Agriculture Reviews, 2, 189–218.

23

Adaptation and Mitigation Strategies of Climate Change for Backyard Poultry

M K Padhi and B C Das

Introduction

The egg production in India is 75 billion and the broiler production is 4.2 million tonnes per annum. The growth rate of layer market is 6-7 percent per annum and broiler market is 8-10 percent per annum. The Indian poultry sector is valued at INR 1 lakh cr or USD 15.38 bn. Approx. 80 percent of egg production is contributed by commercial poultry farms, remaining comes from household/ backyard poultry. Backyard poultry plays a key role in supplementary income generation and family nutrition to the poorest of the poor. It is estimated that there are around 30 million farmers engaged in backyard poultry as per 19th Livestock Census. Backyard poultry are being practiced for its inherent strengths. Free range birds reared by the farmers have the advantage of small body size, coloured plumage, broodiness to hatch chicks, adaptation to harsh climatic condition, lower diseases and production with least investment. The birds are scavenging in the backyard, supplemented by kitchen waste and are kept in low height houses for shelter using locally available materials. It helps to improve the nutritional status of rural poor and also improve the subsidiary income of the poor people. Such stocks make minimal use of land, labour and capital and integrate well into the backyard. Backyard poultry production system is self-sustaining because of minimal needs and can be afforded by poor people of the rural areas. The high yielding white feathered chicken varieties being reared in the intensive poultry farming are not suitable for backyard / free range backyard poultry farming due to the adverse climatic and high pathogen load prevailing in such rearing practices. Backyard poultry farming is an in-separable part of the poor tribal people and it will continue to be due to its advantages. Although backyard poultry production is not an occupation per se, it often supplements household income, primarily for poor families. Strengthening backyard poultry production may help overcome poverty and malnutrition in impoverished communities. However, the effect of climate change on these birds needs to be studied for its sustenance.

Climate may be defined as the weather conditions prevailing in an area in general or over a long period. Climate change occurs when changes in Earth›s climate system result in new weather patterns that last for at least a few decades, and maybe for millions of years. Climate change, caused by emissions from industries and other human activity, is making the world warmer, disrupting rainfall patterns and increasing the frequency of extreme weather events. No country is immune to these forces, but India is particularly vulnerable. Extreme events may be the most tangible and immediate impact of climate change, but another more long-term and equally dangerous effect is rising temperatures. In India, according to IMD data released by the statistics ministry, average temperatures have increased by 0.6 degrees Celsius (° C) between 1901-10 and 2009-18. At an annual level, this may seem trivial, but projections deeper into the future paint a more alarming picture. For instance, the World Bank estimates that, if climate change continues unhindered, then average temperatures in India could reach as high as 29.1° C by the end of the century (up from 25.1° C currently). A primary channel for the fall in incomes comes from climate change effects on farmers. The monsoon and suitable temperatures are critical inputs for farmers. Hotter weather and disrupted rainfall hurt crop yields and, consequently, their incomes (Padmanabhan *et al.* 2019). Poor farmers will be more vulnerable as they are having some agriculture and livestock for their livelihood. Poorest of poor are having backyard poultry as a subsidiary income as well as to feeds its family nutritious eggs and poultry meat. The climate change also affects the productivity of these birds and the society has to take care for their adaptation and mitigation strategies to minimize the impact.

Effect of climate change on backyard poultry

Change climatic situation produce climatic stress. Any deviation from normal condition is called stress. Birds have limited body resources for growth, reproduction, response to environmental changes and defence mechanism. Under stress condition there is redistribution of body resources including energy and protein at the cost of decreased growth, reproduction and health. Under long term stress or repeated stress birds become fatigue or weak. These conditions lead the birds to starvation and infectious diseases. Poultry are subject to frequent stress factors and therefore it is important to have an effective management programme to minimize their effect on performance and health of the birds. There are many common sources of stress. High ambient temperature in the tropics like that of our India accompanied by high relative humidity is one of the most important stress and birds are more susceptible to high temperature than low environmental temperature. High environmental temperature tremendously damages the egg quality. Some precaution may be followed against climatic stress and emphasis to be given to heat stress for improvement of poultry production

Heat stress causes production losses in the intensive poultry production industry. Several reviews have explored possible impacts of a hot climate on poultry production systems – focusing on the physiology and detrimental effects of heat stress (Bhadauria *et al.*, 2014; Lara & Rostagno, 2013; Tankson *et al.* 2001; Winsemius *et al.*, 2014). From such studies, it is evident that high temperatures impose adverse consequences on performance and productivity of commercial poultry. However, there is research gap on how heat stress is impacting backyard poultry production in particular, especially given the importance of such birds to improve food security in poor households (Nyoni *et al.* 2018). In commercial production systems where environmental microclimates are well controlled, local climate (or climate change) may not be of significant concern. However, in backyard areas of developing countries, it is likely that backyard poultry is adversely affected (i.e. stressed) by extreme environmental conditions, as birds continue to interact with the local environment during scavenging, and farmers have less capacity to control their living environments (Winsemius *et al.* 2014). It is opined that increases in temperatures are likely to cause heat stress on poultry especially in rural settings where there is limited study on how environmental conditions impact poultry, and consequently on the associated best management practices. It is thus important to begin assessing the dynamics of backyard poultry production in a warming climate. Several biotic factors such as a species' genetic potential, life stage and nutritional status, may determine the level of vulnerability to heat stress (Thornton *et al.*, 2009). According to Huey *et al.* (2012), species with high physiological capacities to buffer environmental variations and high potential for rapid evolution are better equipped to survive rapid warming. The level of tolerance to high environmental temperatures is also determined by the species' place of origin. Animals that have not been exposed to high temperatures are generally more vulnerable to heat stress than those that have adapted (acclimatized) over time (Sirohi & Michaelowa, 2007).

Since backyard poultry is characterized by slow growth rates, it is possible that they are less susceptible to the impact of heat stress. Research suggests that all birds generally react in a similar manner to heat stress, but may express individual variation of intensity and duration of responses (Lara & Rostagno, 2013). Findings from Malaysia concluded that the susceptibility of birds to heat stress is not determined by genetic differences in body size and age, but rather by domestication and selective breeding (Soleimani *et al.* 2011). Although backyard poultry is generally known to be hardy and well adapted to harsh, stressful environments, scientific information that backs these claims is limited. Since backyard poultry is usually exposed to prolonged high temperatures (in hot regions) during scavenging, it is likely that they have genetically adapted to a warming climate. However, both the impact of increased frequencies of heat waves and critical heat stress threshold levels on backyard poultry has not been scientifically established. It is likely that as backyard poultry dissipate excess heat, they may pant excessively, especially during summer when highest ambient temperatures are experienced. Under this premise, it is possible that the observed low poultry output (in terms of eggs produced and number of birds consumed and/or sold) in some small-scale backyard poultry farming regions may, in part, be a function of heat stress. However, the extent and variation of temperature tolerance thresholds in backyard poultry remains unknown, and is thus a knowledge gap that requires urgent attention (Nyoni 2018).

Adaptation and mitigation strategy

Genetics and breeding: One of the most effective ways of improving heat tolerance/ temperature modulation is through the incorporation of single genes that reduce or modify feathering, such as those for naked neck (Na), frizzle (F) and scaleless (Sc), as well as the autosomal and sex-linked dwarfism genes, which reduce body size. ii. One of the major breeding strategies in India is based on crossing Aseel breed males with CARI Red hens to produce crossbred CARI Nirbheek hens. Kadaknath is an indigenous breed whose flesh is black and is considered not only a delicacy but also of medicinal value and their crosses like CARI-Shyama are also popular. Some of the stocks developed for the purpose are Chabro, Kalinga Brown, Kaveri, Vanaraja, Gramapriya, CARI-Gold, Hitcari, Upcari, Cari-Debendra, Giriraja, Girirani, Krishipriya, Swarnadhara, Nandanam 99 and Rajasri. iii. New varieties have also up like Srinidhi, Jharsim, Kamrupa and Pratapdhan. A few private sector players like Keggfarms, New Dr. Yashwant Agritech Pvt Ltd, Jalgaon, Indbro Research and Breeding Pvt. Ltd., Shipra Hatcheries, Patna are also producing stocks like Kuroiler, Satpuda-desi, Rainbow Rooster and Shipra in this segment.

Breeds/ strains having high immune competence will be another priority area for research due to adaptability of future stocks to changing farming systems and climate. For smallholder systems, nucleus crossbreeding, community based breeding programs and strategies to generate sustained replacement stocks in systems where crossbreds are the best option may be explored further. Selection of indigenous breeds in high temperature similar to climatic condition may be helps the birds to adopt climatic change with improve productivity.

Supplementary feeds if any to be given early in the morning

1. ***Lighting system:*** Switched on in the early morning coolers hours so that the birds starts feeding early. Light 14hrs/day may be given.
2. ***Calcium:*** It is the main mineral component use for the egg production to maintain normal egg shell quality under heat stress. Under heat the backyard poultry may not gate the required nutrient in the backyard as in hot weather everything become dry so they need supplemental calcium feeding during this season,
3. ***Phosphorous:*** The requirement of this element is increased at high temperature.
4. ***Water:*** Under normal ambient temperature condition a layer in full production consume about 200 ml of water however if the temperature increases more than 32^0 C water consumption increase upto 50 % . Adequate cool water may be given to the birds during heat spell.
5. ***Vitamins:*** High temperature may also results in lower utilization of vitamin and certain vitamins in feed are less stable in high temperature. Studies showed that beneficial effect by feeding of Vitamin E in additional quantity to laying hens under heat stress. It is suggested that 250 mg vitamin E is optimal for alleviating at least in part the adverse effect of chronic heat stress in laying hens (Chung *et al.* 2005). Chung *et al.* (2005) also recommend 200 mg vitamin C/kg body weight for heat stressed hen.
6. ***Hormone:*** Exogenous estrogen with high level of dietary Vitamin D, both before heat stress is effective in alleviating at least some effects of heat stress during high ambient temperature (Hansen *et al.* 2004). Lower level of T_3 and T_4 ratio also support the bird to survive under heat stress.
7. ***Density:*** The effect of climatic change leading to heat stress aggravated by increased bird density, feed and water deprivation, inadequate ventilation, vaccine reaction and presence of disease or parasite
8. ***Ventilation:*** Better ventilation is required in the shelter of the backyard poultry
9. ***Grass cover and shed tree:*** Grass cover on the ground surrounding the poultry houses will reduce the reflection of sunlight in to the house. Sheds tree should be located where they do not restrict air movement.
10. ***Aspirin:*** feeding aspirin may reduce body temperature of birds under heat stress.
11. ***Feeding of catecholamine antagonist:*** Chemical agents which block the catecholamine synthesis in the body reduced the heat stress of the

bird and improve the feed intake, quality and quantity of eggs of the bird.

12. ***Humidity:*** Optimum humidity for poultry production is 55 to 65 % (RH).
13. ***Immune system:*** Immune system of the bird is adversely affected by heat stress. Therefore administration of vaccines to birds is not usually recommended during very hot weather.
14. ***Need good indicator of physiological stress:*** the physiological indicator of stress such as atrophy of thymus and bursa of fabrics in young birds, enlargement of anterior pituitary and adrenal glands are good indicator of stress but inherent problems with their detection these organs cannot be weighed in live birds and require slaughter of animal. Therefore suitable technique of physiological indicator of stress is needed.

Physiological stress indicators

1. ***Physical response:*** RT and RR are good indicators of heat stress. Increased RR that is manifested is the first reaction when birds are exposed to environmental temperature.
2. ***Cardiac Function:*** Blood pressure and heart rate reflects the cardiovascular function. Exposure to high ambient temperature (HAT) is associated with a decline in blood pressure, an increase in cardiac output and a decrease in peripheral resistance (Sturkie, 1967), by reducing blood flow to the viscera by 44%. As birds become acclimatized to elevated ambient temperature, cardiac output decreases, blood pressure increases and peripheral resistance returns to normal (Sturkie, 1967). It is also observed that an increase in arterial blood pressure with simultaneous reduction in HR accompanied by decreased heart weight indicates the acquisition of resistance to higher temperatures.
3. ***Haemogrgm and Leukogram profile:*** Reduction in erythrocyte number, hemoglobin content, hematocrit (Borges et 2004; Zulkifi *et al.*, 2009) circulating leucocytes are reported as haematological alterations in response to stress in chicken.
4. ***H/L Ratio:*** Heterophil to lymphocyte (H/L) ratio has been accepted as the sensitive and reliable index for determining the effect of various stressors in poultry (Minka and Ayo, 2011) as under stress conditions. Thermal stress decrease both humoral and cellular immune responses (Kannan *et al.*, 2002), which results in an increase in H : L due to reduced number of circulating lymphocytes and higher numbers of heterophils.

5. ***Biochemical parameters:*** Blood glucose tends to decline with long-term heat exposure (Zulkifli et 2009). Plasma proteins (Tollba and Sabry, 2004), uric acid (Sun *et al.*, 2015) decrease in heat stressed chicken, Albumin, rather than total protein or globulin, has been shown to be the most sensitive indicator of protein status (Aksit et at, 2006), Serum total cholesterol and LDL cholesterol increase and cholesterol decrease due to heat exposure which was attributed to depressed level of thermogenic hormones in heat stress (Habibian et at, 2014). Heat Stress also decreases blood Na, K and partial pressure of carbon dioxide (pC02), which may disturb acid-base balance and cause respiratory alkalosis, respectively (Borges *et al.*, 2004).
6. ***Stress hormones:*** Corticosterone is the major glucocorticoid in birds and involved in metabolic rate and heat production. The circulating levels of corticosterone showed a significant rise from 0d to 3d and consequently decreased on 7d (Gogoi, 2016). T_3 could be considered as reliable indicator of long-term heat stress (Tao *et al.*, 2006; Moudgal *et al.*, 2007). Circulating T_3 concentration significantly reduced with linear increase in temperature (Gogoi. 2016). The ability to reduce plasma T_3 concentration, especially during a thermal challenge, suggests an improvement in thermo tolerance (Yahav, 2000). Lower levels and higher ratio of T_4 and T_3 favours the survivability of birds during summer stress (Moudgal *et al.*, 2007).
7. ***Heat shock proteins (HSPs)and factors:*** Upregulation of HSPs synthesis is considered as an endogenous adaptive phenomenon leading to improved tolerance to various stress conditions/factors. Increased HSPs protect cells against the additional stress (Wang and Edens, 1998). Exposure of poultry species to mild stressors over a period of time enhances HSP70 expression, but eventually, the birds become acclimated and no further increase in cellular HSP70 can be demonstrated (Edens *et al.*, 2001).

Conclusion

Backyard poultry is an important component in poultry production for the rural poor for providing supplementary income as well as nutritious egg and meat to the rural poor. The effect of climatic change on backyard poultry and its adaptation and mitigation strategy is important for its long term sustainability. Since indigenous bird are being used for the farming they are more resistance to climatic change as they are well adopted to climatic stress due to natural selection. However, to improve the production from backyard poultry suitable breeding and selection to be followed to develop breeds/variety those are more adopted to climate change. Different major genes available in the indigenous birds may be of great use. Mitigation strategies to minimize the effect of climate change on backyard poultry to be followed strictly so that the production from them should not decreased substantially. Parameters to measure the stress due to climate change to be studied and suitable indicator should be refined so that it can be measure easily. The same traits may be use in breeding and selection programme to develop a bird adapted to climatic change. Though backyard poultry are more adapted to climatic stress but different mitigation strategies to be put on use whenever the bird is on stress to minimize the effect of stress.

References

Aksit, M. 2006. Poultry Science, 85: 1867-74.

Bhadauria, P. *et al.* (2014). Journal of Poultry Science and Technology, 2(4), 43–56.

Borges, S.A. *et al.* 2004. Poultry Science, 83(9): 1551-8.

Chung *et al* 2005. (Asian-Aust. J. Anim. Sci. 18(4): 545-551.

Edens, F.W., *et al.* 2001. In: Proceedings for the 50th Western Poultry Disease Conference. University of California, Davis California.

Gogoi, S.2016. Ph.D thesis, Indian Veterinary Research Institute, Izatnagar, India.

Habibian, M., *et al.* 2014. International Journal of Biometeorology, 58(5):741-52

Hansen, K *et al.* 2004.. Poultry science. 83. 895-900.

Huey, R. B. *et al.* (2012). Philosophical Transactions of the Royal Society B: Biological Sciences, 367(1596), 1665–1679.

Kannan. G., *et al.* 2002. Journal of Animal Science. 80: 1771-1780.

Lara, L.J. and Rostagno, M. H. (2013). Animals, 3, 356–369.

Minka. N.S. and Ayo. J.O. 2011 ISRN Veterinary Science.

Moudgal. R.P., *et al.* 2007. Indian Journal of Poultry Science 42(1):1-4.

Nyoni *et al.* 2018. Climate and development, https://doi.org/10.1080/17565529.2018.1442792

Padmanabhan, V *et al.* (2019) https://www.livemint.com/news/india/the-growing-threat-of-climate-change-in-india-1563716968468.html.

Sirohi, S., & Michaelowa, A. (2007). Climatic Change, 85, 285–298.

Soleimani, A. F. *et al.* (2011). Poultry Science, 90, 1435–1440.

Sturkie. P.D. 1967. Journal of Applied Physiology. 22: 13–15.

Tankson, J. D. *et al.* (2001). Poultry Science, 80, 1384–1389

Tao. X., *et al.* 2006. Poultry Science. 85: 1520-1528

Thornton, P. K., *et al.* (2009). Agricultural Systems, 101(3), 113–127.

Tollba. A.A. and Sabry M.M. 2004. Poultry Science.24: 333-349.

Wang. S. and Eden's. F.W.1998. Poultry Science. 77: 1636-1645

Winsemius, H. C. *et al.* (2014). Hydrology and earth system sciences discussions. European Geosciences Union,10, 14747–14782.

Yahav. S. 2000. Avian Poultry Biology Reviews. 11: 81-95.

Zulkifli, I., *et al.* 2009. Poultry Science, 88:471-476..

24

Nutritional and Management Practices of Poultry Under Extreme Hot and Humid Condition

A K Panda and B C Das

Introduction

There is an ongoing debate on global warming worldwide as it is seriously impacting climate change. The global temperature has been increased by 0.74±0.18°C during the last century and the climate model projections summarized in the Intergovernmental Panel on Climate Change (IPCC) report indicate that the global surface temperature will probably rise a further 1.1 to 6.4°C during the twenty-first century (IPCC, 2007). High ambient temperature is one of the most important stressors to poultry, which has a direct relationship on profitability of chicken meat and egg production. A seasonal problem in many parts of the world, high environmental temperature (35-43°C in tropical countries) causes economic losses through reducing feed intake while decreasing nutrient utilization, live weight gain, egg production, egg quality and feed efficiency in poultry (Corzo *et al.*, 2003; Panda *et al.*, 2007a; 2008). The ideal temperature for broilers is 10-22°C for optimum body weight and 15-27°C for feed efficiency (NRC, 1994). Layers will produce egg constantly in the temperature range of 10-30°C. Above 30°C, performance in terms of growth and egg production declines.

Zone of thermo neutrality

Poultry is homotherm and thus maintain same body temperature irrespective of the surrounding temperature. In a narrow environmental temperature zone known as thermo neutral zone, the energy required for essential physiological reaction to survive (basal metabolism) is minimum. In the thermo neutral zone, the heat production is equals to heat loss. At temperature above the thermo-neutral zone, birds also have to increase the additional energy produced dissipated as heat. So birds produce additional energy which is dissipated as heat. The thermo-neutral zone decreases as the age of the bird increases.

Table 1: Thermo-neutral zone (°C) for chicken

Age (week)	Broiler	Pullet	Layer
1.	32-33	34-35	
2.	30-31	31-32	
3.	27-29	28-30	
4.	24-27	27-29	
5.	18-22	26-28	
6.		25-27	
7.			
8.			19-22

Stress Physiology

The lack of sweat glands and relatively high body temperature make poultry, especially meat type fast growing chickens, compared with other species, more susceptible to heat stress. During exposure to high ambient temperature, poultry face difficulties in maintaining body temperature. As they do not possess sweat glands, therefore the primary means of heat loss above upper critical temperature is by increasing the breathing rate (panting) and by raising their wings (to increase the surface area). Under stress conditions avian blood undergo a change from acid base balance to alkaline balance (pH 7.44). There is a drop in plasma, decreased level of vitamin C in adrenal cortex, reduction in lymphocyte and depression of immune response. As the temperature rises the birds undergo many changes i.e. rise in temperature increases water consumption, respiration rate, body temperature, inferior egg quality and susceptibility to diseases.

In general, the response of the bird to the stressors is in 3 stages, depending on the nature and duration of stressor. The whole sequence of the events is called general adaptation syndrome.

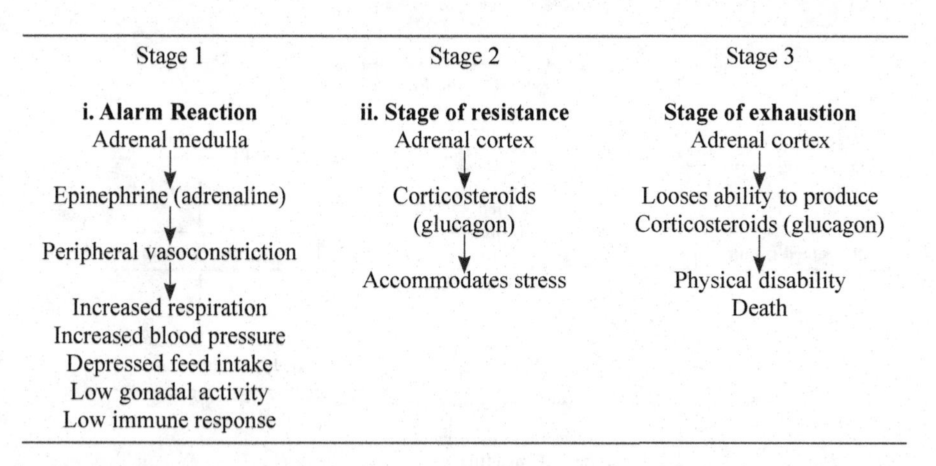

Effect of heat stress in poultry

The most important factor defending the response at high temperature is feed intake. The ambient temperature has a considerable impact on feed consumption of bird especially in adult birds because feed intake decreases as environmental temperature increases. At temperature above 30^{0}c the feed consumption decreases by 2.5 to 4 g per each degree rise in temperature (NRC, 1994). During hot weathers birds reduce feed consumption resulting in some nutrients becoming deficient resulting in poor growth and efficiency. Heat stress is a problem with broilers from 4 weeks onwards and with layers and breeders in production. The effect of heat stress in poultry varied and range from decreased feed intake (NRC, 1994; Yahav, 2000) reduce body weight gain in broilers (Zhang *et al.*, 2012) reduce egg production in layers (Deng *et al.*, 2012) and high mortality (Yahav, 2000).

High temperature: Panting (Hyperventilation) Loss of metabolic CO2 Hypobicarbonaemia Reduced buffering capacity of H^+ during egg shell formation Reduced concentration of carbonic anhydrase Impairment of production of bicarbonate ion in uterus **Reduced shell quality**

High temperature: Reduced feed intake Bone resorption Hyperphosphataemia Reduced capacity of uterus to form calcium carbonate **Reduced shell quality**

High temperature: Demand to regulate body temperature More blood flow to periphery of body Reduced blood flow to uterus **Reduced shell quality**

Stress and immunity

Stressed birds Regression of lymphoid organs (bursa, spleen and thymus) Increasing the ratio of heterophils and lymphocytes in blood

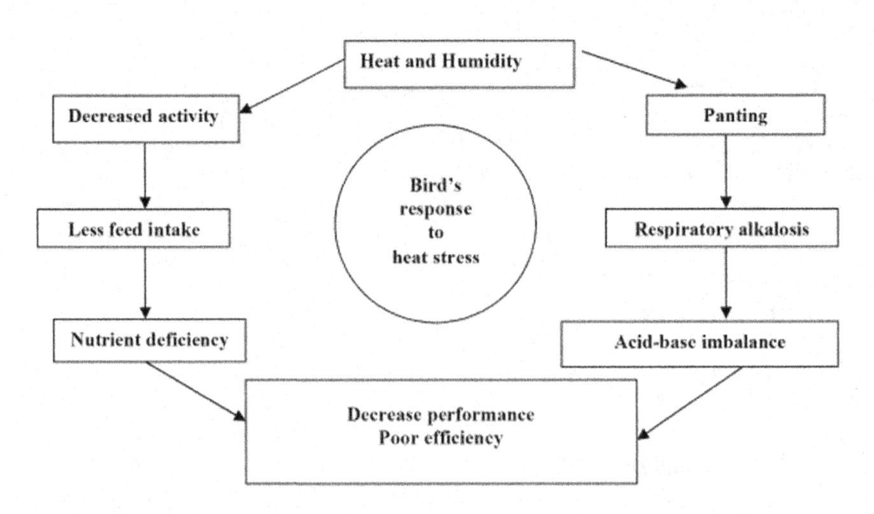

Fig. 1: Bird's response to heat stress

Strategies to alleviate heat stress

Several methods are available to alleviate the negative effects of high environmental temperature on the performance of poultry. The heat stress can be regulated by controlling the environment, where the poultry are being reared i.e. housing temperature and ventilation. However, it is not possible always, because of economic factors. In such cases, the options being left are alternate strategies. One such strategy is nutritional manipulation. Nutritional modification usually involves optimizing the diet to meet the altered needs of stressed birds for energy and protein and providing certain additional nutrients which have shown to have specific beneficial effect (Panda *et al.*, 2009).

Increase nutrient density of the diet

When the environmental temperature increases, feed consumption decreases. The ideal temperature for chicken is 20 to 25^{0}c and temperature in excess of 32^{0}c will lead to heat stress. The higher the humidity, the worse are the effect of heat stress. The reduction in feed consumption changes the dietary levels of energy and protein. The most important factor limiting the bird's performance at high temperature is energy intake. When the environmental temperature increases above 21°C, the energy requirement for maintenance decreases by 30 kcal/day. Although the energy requirement for maintenance is lower at higher temperature, most of the energy is wasted in heat dissipation so the absolute energy requirement is not affected by heat stress. Hence, it is advisable to formulate higher density diet in order to maintain daily intake in line with the requirements for growth and egg production. Use of supplemental fat is suggested as it not only increases palatability of feed but also reduces the amount of heat increment that is produced during its utilization by the body (Zulkifli *et al.*, 2007; Ghazalah *et al.*, 2008).

Protein and amino acids

As the protein and amino acid requirements are independent of environmental temperature, once the protein requirement is met, heat stress does not affect the bird's performance. Protein intake above the requirements or feeding the diets with imbalanced amino acid increase protein catabolism and the associated heat production can markedly increase the heat stress on birds kept at high ambient temperature. Reduction in dietary protein with suitable supplementation of essential amino acids is also a way of reducing heat increment. Therefore, it is advisable to reduce the crude protein content of the diet and supplement the diet with essential amino acids as required (Panda *et al.*, 2007b). Supplementation of methionine hydroxyl analogue rather than DL-Methionine is beneficial in heat stressed bird as it readily absorbed throughout the intestinal tract by passive diffusion, which is a non-energy consuming process (Panda *et al.*, 2007c). To ensure that layers do not suffer nutritional stress of hot weather, it is recommended that protein content of feed should be increased from 16% to 17-18%. By increasing the dietary protein content, the requirements for isoleucine and typtophan could be met, while methionine and lysine can be supplemented with synthetic compounds. There is, however, fear that increasing dietary protein might be detrimental to the bird as more heat is produced during its utilization that may well overload heat dissipation mechanisms (i.e., panting, blood circulation etc.). Therefore, improving overall balance of the diet by amino acid supplementation appears to be more effective than increasing protein intake (Brake etal., 1998; Corzo *et al.*, 2003).

Feeding calcium carbonate or oyster shells

One of the important aspects is to adjust the calcium content of the diet such that each bird consumes the right amount per day. For laying hens, top dressing feed with oyster shell or large particulate limestone is beneficial and has the added advantage of stimulating feed consumption. Limestone and oyster chips may be provided at a rate of 500-600 g per 100 hens.

Supplementation of vitamins and minerals

Additional allowances of ascorbic acid (vitamin C), Vit A, E and D_3 can improve bird performance at high temperature. High temperature also affects metabolism and synthesis resulting in requirement of nutrients such as vitamin

C that are considered essential under normal conditions. Supplementation of diet with 200 ppm of vitamin C has been shown during the heat stress (Panda *et al.*, 2007a). Biotin may be added to reduce the metabolic disorders like fatty liver and kidney syndrome during summer. Addition of water soluble vitamins is beneficial during heat stress. Heat stress is known to interfere with the conversion of vitamin D_3 to its metabolically active form, so higher level may be supplied in the diet.

In laying hens vitamin E supplementation at 125-250mg/kg improves egg production, feed efficiency and immunecompetence. Heat stress depresses egg production due to lower plasma concentrations of egg yolk precursors, vitellogenin and very low density lipoprotein. Vitamin E improves egg production by facilitating the release of vitellogenin from liver and increasing the circulating supply of this precursor for yolk formation by protecting the liver from lipid peroxidation and damage to cell membranes.

Table 2: Effect of dietary vitamin E supplementation on production performance of White Leghorn layers under tropical summer conditions (Panda et al., 2008)

Vitamin E (mg/kg)	Production performance		Immune response	
	Egg production (%)	Food conversion (g egg mass/ g feed)	Newcastle disease virus titre (ELISA titre)	Lymphocyte proliferation ratio
25	84.78[b]	0.441[b]	3275[b]	0.537[b]
125	85.99[a]	0.452[a]	3326[a]	0.549[a]
250	86.03[a]	0.451[a]	3332[a]	0.551[a]

Mineral balance particularly Ca and P are important. It is advisable to increase the available P content of the feed. General survivability to high temperature was observed in birds given low Ca and high P then given normal Ca and low P. Heat stress reduces Ca intake and the conversion of vitamin D_3 to its metabolically active form. Therefore, in laying birds Ca should be provided in the required quantity to maintain the egg shell quality. Supplementation of chromium (Cr) to broiler diet is an effective method to decrease heat stress in broiler. Cr can be added @ 10-20 ppm in broiler ration to reduce the heat stress.

Electrolytes/Buffering agents

Imbalances in acid-base balance occur in during heat stress. Therefore, inclusion of various compounds (sodium bicarbonate, potassium chloride, chloride, ammonium chloride and ascorbic acid) in the diet or water is a common practice to alleviate the adverse effects of heat stress (Ahmad *et al.*, 2005). With rise in environmental temperature, bird's respiration rate increases in an attempt to increase the rate of evaporative cooling. Such panting increase CO_2 loss and alkalosis develops. Such changes in electrolytes balance may result in reduced growth rate in broilers and decline egg shell quality often seen in high producing layers. Prevention of electrolyte imbalance should be obviously tried through incorporation of cation and anion in diet formulations. To mitigate the heat stress, dietary allowances for electrolytes such as sodium, potassium and chloride may be increased by 1.5% for each degree rise in temperature above 20°C.Results of the research findings revealed that dietary supplementation of sodium bicarbonate in the diet or drinking water of heat stressed broilers resulted in improved feed intake, growth and liveability and egg weight and shell thickness in layers. Supplementing the diet with 0.5% sodium bicarbonate or 0.3 to 1.0 % ammonium chloride can alleviate the alkalosis caused by heat stress.

Feed restriction

Under extreme conditions of heat stress, withholding feed can prevent high mortality. Restriction is beneficial as body temperature rises 7 to 12 % two hours after feeding. Feed restriction may reduce the gain but significantly increases the survivability under extreme hot conditions. Feed restriction may be done at least 3 hours before stress, as restriction is not beneficial when the bird is on stress or panting.

Feeding practices

Reduced feed consumption is the main cause of poor performance at high temperatures and the feeding practices suggested below are reported to improve performance of birds under heat stress. The poor performances of the birds at high temperature are due to low feed consumption. So in order to increase the feed consumption the following practices should be encouraged.

- Increase frequency of feeding
- *Form of feed:* Ensure good physical quality of feed (crumb, pellets or mash) to encourage appetite. If there is enough floor space, extra feeders should be added.
- *Time of feeding:* Encourage eating at cooler times of the day, i.e., early morning or in the evening. Feeding birds at cool times enables birds to make up for what they have not eaten during the day. Laying hens increase their calcium intake during the evening as eggshells are normally formed during this time.

- *Wet mash feeding:* Wet mash feeding results in increased feed intake and improved performance mediated by more water intake. But it has certain limitations.
- *Feed storage:* Feed should not be stored for longer than two months, especially in summer to reduce the possibility of mycotoxin build up.
- *Lighting dim the lights while feeding:* using low light intensity during periodic feeding reduces activity that reduces heat load.
- *Addition of fat/molasses:* It will increase the palatability of feed and stimulates the feed consumption.
- *Dietary supplementation with additives:* It is well known that heat stress even affect the nutrient digestibility and absorption, since it alters the structure of intestinal cells and imbalances the normal flora of the intestines. Under these conditions, supplementation of diet with Lactobacillus and Streptococcus is found to be beneficial.
- *Chemotherapeutic agents:*Large no of compounds are available to reduce the stress associated with hypothermia. Aureomycin has been found to alleviate growth depression caused due to stress. Resinpine, an alkaloid is known to prevent loss of CO_2 from birds subjected to high environmental temperatures, thereby stabilizing the acid-base balance in the blood.

Conclusion

High environmental temperature leads to deficiency in specific nutrients because of decline in feed intake, reduced absorption and poor utilization of vitamins. Several methods are available to alleviate the negative effects of high environmental temperature on the performance of poultry. The heat stress can be regulated by controlling the environment, where the poultry are being reared i.e. housing temperature and ventilation. Nutritional modification usually involves optimizing the diet to meet the altered needs of stressed birds for major nutrients and providing certain additional nutrients which have shown to have specific beneficial effect.

References

Ahmad, T., Sarwar, M., Mahr-Un-Nisa, Ahsan-Ul-Haq and Zia-Ul-Hassan, 2005. Influence of varying sources of dietary electrolytes on the performance of broiler reared in a high temperature environment. Animal Feed Science and Technology, 20: 277-298.

Brake, J., D. Balnave and J.J. Dibner, 1998. Optimum dietary arginine: lysine ratio for broiler chickens is altered during heat stress in association with changes in intestinal uptake and dietary sodium chloride. British Poultry Science, 39: 639-647.

Corzo, A., E.T. Moran and D. Hoehler, 2003. Lysine needs of summer-reared male broilers from six to eight weeks of age. Poultry Science, 82: 1602-1607.

Deng, W., X.F. Dong, J.M. Tong and Q. Zhang, 2012. The probiotic *Bacillus licheniformis* ameliorates heat stress-induced impairment of egg production, gut morphology and intestinal mucosal immunity in laying hens. Poultry Science, 91: 575-582.

Ghazalah, A.A., M.O. Abd-Elsamee and A.M. Ali, 2008. Influence of dietary energy and poultry fat on the response of broiler chicks to heat stress. International Journal of Poultry Sciences, 7: 355-359.

Intergovernmental Panel on Climate Change (IPCC) 2007. Climate Change 2007: The Pysical Science Basis. Contribution of Working Group I to the Fourth Assessment Report of the Intergovernmental Panel on Climate Change. http://ipccwg1.ucar.edu/wg1/Report/AR4WG1_Print_SPM.pdf.

NRC. 1994. Nutrient Requirements of Poultry. 9th rev. ed. National Academy Press, Washington, DC.

Panda, A. K., S. V. Rama Rao and M. V. L. N. Raju. 2007a. Effect of vitamin C supplementation on performance, immune response and antioxidant status of heat stressed White Leghorn layers. Indian Journal of Poultry Sciences, 42: 169-173.

Panda, A. K., S. V. Rama Rao, M. V. L. N. Raju, G. Shyam Sunder, R. N. Chatterjee and R. P. Sharma. 2007b. Effect of Supplemental DL-Methionine on Performance of Commercial Laying Hens during summer. Animal Nutrition and Feed Technology 7: 169-175.

Panda, A. K., S. V. Rama Rao, M. V. L. N. Raju and S. K. Bhanja. 2007c. Relative performance and immune response in White Leghorn Layers fed liquid DL-methionine analogue and DL-methionine. Asian-Australian Journal of Animal Sciences 20(6): 948-953.

Panda, A. K., S. V. Rama Rao, M. V. L. N. Raju and R. N. Chatterjee. 2008. Effect of vitamins E and C supplementation on production performance, immune response and antioxidant status of White Leghorn layers during summer stress. British Poultry Science, 49: 592-599.

Panda, A. K. 2009. Nutritional management of heat stress in poultry. Asian Poultry, 44-46, October 2009

Yahav, S., 2000. Relative humidity at moderate temperature: its effect on male broiler chickens and turkey. British Poultry Science, 41: 94-100.

Zhang, Z.Y., G.Q. Jia, J.J. Zuo, Y. Zhang, J. Lei, L. Ren and D.Y. Feng, 2012. Effects of constant and cyclic heat stress on muscle metabolism and meat quality of broiler breast fillet and thigh meat. Poult. Sci., 91: 2931-2937.

Zulkifli, I., Htim Nwe Nwe, A.R. Alimon, T.C. Loh and M.Hair-Bejo, 2007. Dietary selection of fat by heat-stressed broiler chickens. Asian-Australian J. Anim. Sci., 20: 245-251.

25

Identification of HSP-70 Gene in Ganjam Goat: DNA Extraction, PCR and Electrophoresis

Chinmoy Mishra and G D Nayak

Blood sample collection

Blood samples (5 ml per goat) will be collected aseptically from the jugular vein in a vacutainer tube (B.D. Bioscience, Germany) containing EDTA as anticoagulant and transferred to the laboratory in an ice box.

Isolation of genomic DNA

The Genomic DNA will be isolated from the blood samples by 'Phenol: chloroform isolation' method as described by Sambrook and Russel (2001). The steps for isolation of genomic DNA are as follows:

- The blood samples will be taken out of deep freeze and will be thawed at room temperature.
- Thawed blood samples will be transferred to 15 ml polypropylene centrifuge tube and will be centrifuged @3000 rpm for 20 min at room temperature.
- The reddish tinged supernatant, containing plasma and lysed RBC will be discarded by careful pipetting.
- The pellet containing WBC and RBC will be mixed with more than 2 volume of chilled RBC lysis buffer and kept on ice for 10 min after gently mixing it end to end, once or twice.
- It will then again centrifuged @ 3000 rpm for 15 min at room temperature and the black tarry colored supernatant, containing lysed RBC will be discarded by pipetting.
- Steps 4 and 5 will be repeated 3-4 times till the WBC pellet became free of the reddish tinge.

- Once a clear 'Off white' colored pellet of WBC is obtained, DNA extraction buffer will be added @ 3 ml per 10 ml blood and vortexed gently to disperse the WBC pellet in the extraction buffer.
- WBC pellet mixed with DNA extraction buffer will be incubated at 37°C for 30 min.
- Subsequently, 10% SDS @ 200 ml per 10 ml blood will be added and mixed gently by inverting the tube once or twice. Care should be taken while mixing, because after adding SDS, lysis of cell wall occurs and DNA now lies fully exposed. As a result the contents of the tube now become viscous.
- Finally, proteinase-K (20 mg/ml solution) will be added @ 20 ml/10 ml of blood, in two pulses. Half of the requirement will be added in first pulse, after which content of the tube will be mixed gently, end-to-end and incubated at 50°C for 3-4 h. This will be followed by the second pulse of remaining amount and overnight incubation at 50°C.
- On the 2nd day, equal volume of Tris-saturated phenol (pH >7.8) will be added to the tube.
- The tubes will be kept on shaker and gently rotatory movements will be given for 15 min to mix the contents thoroughly.
- Subsequently, the tubes will be centrifuged @ 4000 rpm for 20 min at room temperature.
- After centrifugation, the contents of the tubes gets separated into two layers, the upper aqueous layer (containing DNA) and a lower heavier layer of phenol (containing proteins). A white, thin layer, of protein will be also visible at the interphase of the two layers; the upper aqueous phase will be transferred to another 15 ml polypropylene tube with the help of 1 ml wide bore (3 mm diameter) microtip. Care will be taken not to disturb the protein interface layer.
- Similar extraction (as in steps 11-14) will be done once with Phenol: Chloroform: Isoamyl alcohol (25: 24:1) and once with chloroform: Isoamyl alcohol (24: 1).
- Finally, aqueous phase will be taken in a 15 ml polypropylene centrifuge tube and 3 M sodium acetate @ 100 ml per ml of aqueous phase will be added to it followed by gently mixing.
- More than 2 volumes of chilled isopropanol will be added to the tube and mixed gently by swirling the tube once or twice. The tube will be left at room temperature to allow the precipitation of DNA.

- The precipitated DNA will be transferred into a sterile 1.5 ml eppendorf tube (using the wide bore microtip of 1 ml capacity) along with 500 ml isopropanol and centrifuge 10,000 rpm for 10 min at room temperature.
- Supernatant will be discarded without interfering with the DNA pellet.
- The DNA pellet will be washed twice with 70% ethanol, similar to steps 18 and 19.
- Finally, DNA pellet will be air dried for 1 h to remove traces of ethanol and subsequently dissolved in 200 ml of TE buffer.
- The eppendorf tubes will be kept in water bath at 60°C for 2 h to inhibit DNAse activity and to dissolve pellet properly in the TE buffer.
- After 2 h of incubation, the DNA will be cooled and stored at -20°C for further use.

Evaluation of quality, purity and concentration of DNA

The genomic DNA isolated from the blood samples will be checked for quality, purity and concentration. Only the DNA samples of good quality, purity and concentration will be used for further analysis.

Quality of genomic DNA

Horizontal submarine agarose gel electrophoresis will be performed to check the quality of genomic DNA using 0.8 % (w/v) agarose. At first, the gel casting tray will be prepared by sealing its both ends with adhesive tape and then the comb will be arranged over it in such a way that there remains a gap at least 0.5 mm between the tips of the comb teeth and floor of the casting tray, so that the wells will be completely sealed by agarose.

Subsequently, 0.8% agarose (w/v) suspension in 1X TBE buffer will be heated on an electric heater until the agarose will be completely melted and dissolved to give a clear transparent solution. After cooling it to 60°C, ethidium bromide (10 mg/ml) @ 5 ml per 100 ml of agarose solution will be added to a final concentration of 0.5 mg/ml and will be mixed gently.

The agarose solution will be poured in to the sealed casting tray. The gel will be prepared to about 4 mm thicknesses. The agarose gel will be allowed to set completely at 4°C temperature and then the comb will be gently removed. The adhesive tape will be also removed and gel casting platform was submerged in the electrophoresis tank containing 1 X TBE buffer.

For loading the samples, 10 ml autoclaved triple distilled water along with 5 ml diluted DNA will be taken and after mixing it with 1 ml of 6 X gel loading dye (xylene cyanol and bromophenol blue), it will be loaded in to the well with the help of microtips. Electrophoresis will be performed at 5V/cm (40-50 V) for 3 h. Once the electrophoresis is over, the gel will be visualized under UV transilluminator and documented by photography under Gel Documentation System. Only DNA samples showing intact bands and devoid of smearing will be used for further analysis.

Purity of genomic DNA

The purity of genomic DNA will be checked using UV-Spectrophotometer. A total of 6 ml of genomic DNA of each sample will be dissolved in 294 ml of triple distilled water and the spectrophotometer reading will be taken at OD_{260} and OD_{280} against 300 ml triple distilled water as blank.

Only the genomic DNA samples lying in the ranges of OD ratio (260:280) between 1.7 to 1.9 is considered as good and will be used for further study (PCR amplification) and those showing value beyond this range will be reprocessed by phenol: chloroform extraction method.

Primer designing for amplification of HSP 70 gene

The goat specific nucleotide sequence for HSP70 gene will be retrieved from NCBI public database. One pair of primer for amplification of HSP70 gene in the goat will be designed using available nucleotide sequence in the NCBI.

Polymerase chain reaction

Gradient PCR will be carried out to optimize the annealing temperature for HSP70 genes using the thermoycler (My Cycler, Bio-Rad). The optimum concentration of PCR reaction components will be standardized. A negative control without template DNA will be run in order to check any contamination.

Agarose gel Electrophoresis

PCR products will be checked by horizontal submarine 2% agarose gel electrophoresis in 1X TBE (Tris-Borate EDTA) buffer along with 100 bp DNA ladder (GeneRuler, Thermo scientific). The electrophoresis will be done at 60 volts for 2 hr. The amplified product will be visualized and documented by gel documentation system. The PCR products will be stored at -20°C till further use.

26

Impact of Climatic Variables on Physiological, Haemato-biochemical Profile of Poultry

Pravas Ranjan Sahoo

Introduction

Poultry production is one of the fastest growing sectors of livestock industry in developing countries. Environmental variation is one of the major factors that affect sustainability of poultry production systems in tropical climate (Sinha *et al.*, 2017a). Adverse climatic stress declines the production performance of chicken which causes high economic loss to the enterprises as well as the farmers. Heat stress has an adverse effect on egg production, egg weight and shell quality of laying hen (Whitehead *et al.*, 1998). Climatic variables like temperature, humidity, radiation and wind speed that directly affect the mechanism of thermoregulation and rates of heat exchange by all animals (NRC, 2001). Heat stress is a major factor that decreases productivity and reproductive efficiency of livestock due to lower feed intake and negative energy balance (De Rensis and Scaramuzzi, 2003). Moreover, it badly affect the production due to decline feed digestibility such as proteins, fats, starch (Bonnet *et al.*, 1997). In addition, the acute heat stress drastically decreases the reproductive performance of hens due to alterations in acid-base balance and ion exchange mechanism (Mahmoud *et al.*, 1996). Climate change influences the emergence of disease and their transmission due to increases vectors, pathogens. All the above effects impact a huge economic loss to the farmers as well as the country.

Sign of heat stress in poultry

The signs of heat stress in poultry are panting with open mouth, elevated their wings and squatting near to the ground, droopy acting, slowness and lethargic closed eyes, lying down, increased water intake, decreased appetite, drop in egg production, reduced egg size, poor egg shell quality, reduced body weight, and increased cannibalism (Dayyani and Bakhtiyari, 2013). Birds are trying to lose heat by gasping and changing the position of their feathers, losing water in their breath and cooling by evaporation through the surface of the lungs. Birds are facing to heat stress conditions, they spend less feeding time during feeding, more time drinking and panting, less time moving or walking and more time resting (Mack *et al.*, 2013).

Effect of climatic variables on poultry production

Climatic variable adversely affects the efficiency of broiler production and their meat quality. High environmental temperature and THI value above the critical thresholds level lead to reduced feed intake, lower body weight, and lower feed conversion efficiency (Sohail *et al.*, 2012). Chronic heat exposure adversely affects the meat quality and fat deposition in broilers, but it is breed dependent. Exposure of bird to high solar radiation is coupled with depression of chemical composition and meat quality in broilers industry. Exposure of high temperature during the growing phase of broilers has been related with poor meat characteristics of broiler chicken and loss their quality. Moreover, exposure of heat stress during transportation of birds from production farms to processing centre affect the meat quality. Exposure of laying hens to climatic stress also resulted in a significant decrease in egg production and egg quality. Various author reported that reducing egg production in hot weather due to decrease in feed intake, reducing the uptake of available nutrients and decreases digestibility of different components of the diet (Allahverdi *et al.*, 2013). So, egg production is inversely correlated with environmental temperature beyond the critical limit. The lowered amount of CO2 in blood causes the rise in blood PH which reduces the level of Ca2+ ion in the blood that utilized by the shell gland results poor egg quality).

Effect of climatic variables on reproduction

Climatic variable has a great impact on the reproductive performance and egg quality of birds. High temperature with high relative humidity has more detrimental effect on reproduction of animal. Exposure of White Leghorn hens to high environmental temperature causes decline in reproductive activity, leading to reproductive failure and poor egg quality (Ebeid *et al.*, 2012). The reduction in reproductive performance of domestic birds coupled with heat stress due to decrease LH levels and hypothalamic gonadotropin releasing hormone-I content. Diminishing reproductive performance of poultry in heat stress due to disturbance the thermoregulatory mechanisms, might be modulated at the level of the hypothalamus and

Effect of climatic variables on behavioral and physiological responses

Thermoregulation is an important role in maintaining the homeostasis and it is controlled by central, metabolic and endocrine systems. The body mass, confirmation and morphological parameters such as fur color are related to basal metabolic rate, can use of behavioral adjustments (Cooper *et al.*, 2008). Thermoregulatory capacities of animal play an adaptive role to survive in adverse environment. Under high environmental temperature, birds change their behavioural and physiological responses to maintain their body temperature through seeking thermoregulation. Birds are subjected to under heat stress conditions time spend less in feeding, more in drinking, panting, and wings elevation, move towards cooler surfaces. In adverse climatic condition, maintaining homeostasis mechanism in birds by heat exchange between environment and air sac through convection, evaporative heat loss, perspiration and vasodilation process

Effect of climatic variables on the immunolgical responses

Heat stress has negative effects on health status of birds leading to changes in physiology, metabolism, hormonal and immune system. At high temperature decreases synthesis of T and B lymphocytes and suppression of phagocytic activity of blood leukocytes. Bartlett and Smith (2003) found that lower levels of total circulating antibodies and lower levels of specific IgM and IgG in broiler under heat stress. It was also reported that decrease total WBC and activities of leukocytes subsequent heat exposure. Zulkifi *et al.*, (2000) also supported that heat stress had significantly decline in antibody production. Inflection of the immune response by the central nervous system (CNS), and is mediated by a complex network of nervous, endocrine and immune systems.

Effect of climatic variables on the biochemical profile of poultry

The blood is functioned as the carrier of nutrients, metabolic wastes and the pathway of humoral transmission. So the blood biochemical parameters would reflect the physiological state of the body (Lin *et al.*, 2000). Previous studies reflect that there was alternation in haematobiochemical parameters with the climatic change. Decreased haematocrit, haematoglobin and total protein by high temperature and also decreased concentrations of plasma free amino acids and essential amino acids and increased uric acid concentration were reported. Increased level of free fatty acids and blood glucose level were found during heat stressed laying hens. Similarly, declining of plasma K and increased level of plasma Cl were also stated during exposure to heat stress. So these findings lead to a detail evaluation of different biochemical parameters releated to climatic change.

Blood Glucose

Glucose is considered as the chief metabolic fuel and is responsible for positive energy balance which indirectly supports the normal growth, maintenance and production of poultry. The blood glucose concentration in poultry varies according to climatic change such as temperature, humidity and also due improper manegemental practices. Blood glucose can be estimated by GOD POD method (Tinder, 1969) with using suitable the reagent kit in colorimeter. The Optical Density is always measured spectrophotometrically at 505 nm wave length to assay the concentration of glucose in mg/ dl.

Total protein and albumin

The steroid hormones of reproductive importance of poultry function through the regulation of protein synthesis. The plasma protein content varies in respect to the hormonal status of the birds in various physiological stages such as hatching and brooding. The serum total protein mainly comprises of albumin and globulin which are exclusively synthesized in liver and maintained in plasma pool to serve as the transporting, buffering and osmolar agent of the liquid connective tissue. The serum total protein and albumin are the reflections of normal liver functions because the major plasma proteins are exclusively synthesized in liver. The concentration of total serum protein and albumin can be estimated by Biuret and Bromo Cresol Green (BCG) method in using standard kit of following the description Gornall *et al.,* 1949 and Rodkey, 1964 respectively. The OD of total protein and albumin are recorded colorimetrically at 545 and 570 nm to assay the concentrations in g/ dl respectively.

Cholesterol

Cholesterol is the synthetic product of liver which indirectly states about the functional status of it through any altered value in serum. Cholesterol is the precursor of steroid hormones and indicates the physiological and reproductive status of the poultry birds because various reproductive stages are regulated by the hormones and under the influence of which the serum cholesterol level varies within birds. The high level of cholesterol is an indicative of more secretion of steroids due to non-specific causes which increases cholesterol turnover from plasma pool. The cholesterol can be estimated in serum by CHOD/ PAP method using the kit by following the description (Allian *et al.*, 1974). The OD is measured spectrophotometrically at 505 nm wave length against the blank within 60 minutes to assay the concentration in mg/ dl.

Urea

Serum urea is the nutritional indicator of nitrogen balance and protein metabolic status where its concentration varies as per the availability and utilization of nitrogen. A positive nitrogen balance facilitates the individual to remain healthy in respect of growth, production and maintenance. Its concentration is influenced by a variety of interrelated parameters including dietary protein intake, degradability, amino acid composition, protein and carbohydrate requirement, liver and kidney functions and muscle tissue breakdown. The serum urea concentration can be estimated in serum by modified method of Berthelot, 1964 and the OD will be measured spectrophotometrically at 570 nm wave length against the blank within 60 minutes to express the conc. as mg/ dl.

Liver Enzymes

Intracellular enzymes like AST, ALT and ALP being the units of cell metabolism, take part in different metabolic path ways of poultry birds and exist either in free/ bound form. The cellular/ tissue/ organic functions and body metabolism are therefore influenced by the enzyme level. Further, the hormone action is carried out through synthesis of these enzymes which in turn regulates the body metabolism. The metabolic status of normal growth, maintenance, production and reproduction are reflected as alterations in serum/ plasma enzyme concentration. Any extent of membrane damage leads to leakage of these enzymes from cell. The extent of leakage correlates linearly to the degree of cell damage and for which the enzymes are used as marker for liver disorders. These enzymes many times are associated with metal ions to form activated enzyme complexes. The serum AST and ALT activities can be assayed by UV-Kinetic method of Reitman and Frankel (1957) at 340 nm wave lengths and the activities are expressed in IU/ L. The serum ALP activity can be assayed by UV-Kinetic method at 405 nm wave length and the activities are expressed in IU/ L.

Serum Catalase (CAT)

It is an important intracellular enzyme found in nearly all living organisms inkling poultry birds which are exposed to oxygen. It catalyzes the decomposition of hydrogen peroxide to water and oxygen, in protecting the cell from oxidative damage by reactive oxygen species (ROS) in birds. The enzyme activity was assessed by measuring catalase degradation of H_2O_2 using a redox dye, according to standard procedure (Cowell *et al.* 1994).

Serum Super oxide dismutase (SOD)

This enzyme that catalyzes the dismutation (or partitioning) of the superoxide (O_2^-) radical into either ordinary molecular oxygen (O_2) or hydrogen peroxide (H_2O_2). Superoxide is produced as a by-product of oxygen metabolism and, if not regulated, causes many types of cell damage. Hydrogen peroxide is also damaging and is degraded by other enzymes such as catalase. Thus, SOD is an important antioxidant defense in nearly all living cells exposed to oxygen. SOD activity was measured by the **xanthine oxidase** method which monitors the inhibition of nitro blue tetrazolium reduction by the sample (Sun *et al.*, 1988).

Serum glutathione peroxidase (GPx)

The biochemical function of glutathione peroxidase is to reduce lipid hydroperoxides to their corresponding alcohols and to reduce free hydrogen peroxide to water. It was assessed by using H_2O_2 and an electron donor dye that forms a pink color during the peroxide reaction according to Kokkinakis and Brooks (1979).

Glutathione-S- transferase (GST)

This enzyme catalyzes the conjugation of the reduced form of glutathione (GSH) to xenobiotic substrates for the purpose of detoxification. This activity was assessed by measuring the conjugation of 1-chloro-2, 4-dinitrobenzene (CDNB) with reduced glutathione according to Habig *et al.* (1974).

Thiobarbituric acid reactive substances (TBARS)

These are formed as a byproduct of lipid peroxidation which can be detected by the TBARS assay using thiobarbituric acid as a reagent. It is one important indicator for the climatic change in poultry, can be assayed by the standard procedure Ohkawa *et al.* (1979)

Uric acid

In poultry birds, uric acid also is the end-product of purine metabolism, but it is excreted in feces as a dry mass. This involves a complex metabolic pathway that is energetically costly in comparison to processing of other nitrogenous wastes such as urea (from the urea cycle) or ammonia, but has the advantages of reducing water loss and preventing dehydration. This method cane be estimated by uricase method

Conclusion

The climatic changes such as thermal stress, oxidative stress as well as environmental pollution are the major issues for the productive performance of the poultry birds not only in India but also in the global scenario. These affect different physiological alterations including increased body core temperature, decreased feed intake, increased thirst, altered blood pH, electrolyte imbalances and impaired reproductive functions etc. This also affects the blood biochemical parameters including blood glucose, total protein, albumin, uric acid, cholesterol, ALT, AST and blood urea level of the birds. These can be minimized by supplementing vitamins, minerals, electrolytes and some managemental aspects other than nutrition such as feeding and lighting practices etc to combat the economic loss in poultry industry.

References

Allahverdi A, Feizi A, Takhtfooladi HA and Nikpiran A. (2013). Effects of Heat Stress on Acid-Base Imbalance, Plasma Calcium Concentration, Egg Production and Egg Quality in Commercial Layers. Global Veterinaria.10 (2): 203-207.

Bartlett JR. and Smith MO. (2003). Effects of different levels of zinc on the performance and immunocompetence of broilers under heat stress. Poult. Sci., 82: 1580–1588.

Berthelot, M.P.E. (1859) Berthelot's Reaction Mechanism. Report de Chimie Applique, 2884

Bonnet, S., Geraert, P.A. Lessire, M. Carre, B. and Guillaumin, S. (1997). Effect of high ambient temperature on feed digestibility in broilers. Poult. Sci., 76: 857–863.

Dayyani, N. and Bakhtiyari, H. (2013). Heat stress in poultry: background and affective factors. International journal of Advanced Biological and Biomedical Research. 1(11): 1409-1413.

De Rensis, F. and Scaramuzzi, R.J. (2003). Heat stress and seasonal effects on reproduction in the dairy cow-a review. Theriogenology. 60: 1139e51.

Ebeid TA, Suzuki T. and Sugiyama T. (2012). High ambient temperature influences eggshell quality and calbindin-D28k localization of eggshell gland and all intestinal segments of laying hens. Poultry Sci., 91: 2282–2287.

Gornall A G, Bardawill C J and David M M. (1949). Determination of serum proteins by means of the biuret reaction. J. Biol. Chem. 177:751-66.

Habig WH, Pabst MJ and Jakoby WB. (1974). Glutathione S-transferases. The first enzymatic step in mercapturic acid formation J Biol Chem. 249(22):7130-9.

Kokkinakis DM and Brooks JL. (1979). Tomato peroxidase: purification, characterization, and catalytic properties Plant Physiol. 63(1):93-9.

Mack LA, Felver-Grant JN, Dennis RL. and Cheng HW. (2013). Genetic variation aiter production and behavioral responses following heat stress in 2 strains of laying hens. Poult. Sci., 92: 285-294.

Mahmoud, K.Z., Beck, M.M., Scheider, S.E., Forman, M.F., Anderson, K.P. and Watchman, S.D. (1996). Acute high environmental temperature and calcium estrogen relationship in the hen. Poult. Sci., 75: 1555-1562.

NRC, (2001). Nutrient requirements of poultry. 9ed National Academy Press, Washington, DC.

Ohkawa H, Ohishi N, Yagi K. (1979). Assay for lipid peroxides in animal tissues by thiobarbituric acid reaction. Anal Biochem. 95(2):351-8

Reitman S and Frankel S. (1957). A Colorimetric Method for the Determination of Serum Glutamic Oxalacetic and Glutamic Pyruvic Transaminases. American Journal of Clinical Pathology. 28 (1): 56–63

Rodkey FL. (1964).Tris (hydroxymethyl) aminomethane as a standard for kjeldahl nitrogen analysisclin chem.10:606-10

Shoal MU, Hume ME, Byrd JA, Nisbet DJ, Ijaz A, Sohail A, Shabbir M.Z. and Rehman, H. (2012). Effect of supplementation of prebiotic mannanoligosaccharides and probiotic mixture on growth performance of broilers subjected to chronic heat stress. Poult. Sci., 91: 2235–2240

Sinha, R., Kamboj M.L., Ranjan, A. and Lathwal, S.S. (2017b). Effect of modified housing on behavioural and physiological responses of crossbred cows in hot humid climate. Indian J. Anim. Sci., 87 (10): 1255–1258.

Sun Y, Oberley LW and Li Y. (1988). A simple method for clinical assay of superoxide dismutase. Clin Chem. 34(3):497-500.

Trinder P. (1969) Determination of blood glucose using an oxidase-peroxidase system with a non-carcinogenic chromogen. J Clin Pathol. Mar;22(2):158-61

Whitehead, C.C., Bollengier-Lee, S., Mitchell, M.A. and Williams, P.E.V. (1998). Vitamin E can alleviate the depression in egg production in heat stressed laying hens. In: Proc. Of spring meeting, WPSA-UK Branch Scarborough. Pp. 55-56.

Zulkifi I, Norma MT, Israf DA. and Omar AR. (2000). The effect of early age feed restriction on subsequent response to high environmental temperatures in female broiler chickens. Poult. Sci., 79:1401–1407.

27

Application of Nanotechnology in Biology and Its Impact on Climate Change

Priyabrat Swain

Introduction

Nanotechnology and nanoparticles are increasingly recognized for their potential applications in various aspects of human animal and animal welfare like development of various healthcare or cosmetic products, nano-electronics and techniques for environmental remediation, and many consumer products. Nanoparticles, by definition, are structures that have one dimension in the 1–100 nm range. Nanotechnology involves the application of materials at the nanoscale to new products or processes. Over the past few decades, inorganic nanoparticles, whose structures exhibit significantly novel and improved physical, chemical, and biological properties, phenomena and functionality due to their nanoscale size, have elicited much interest. Nanophasic and nanostructure materials are attracting a great deal of attention because of their potential for achieving specific processes and selectivity, especially in biological and pharmaceutical applications. Nanotechnology has made a new generation industrial revolution by developing product and formulations for medical, agriculture (nano-fertilizers, nano-herbicides, nano-pesticides, recalcitrant contaminants from water, nano-scale carriers, nanosensors), veterinary care, fisheries and aquaculture, detection of nutrient deficiencies, preservation, photocatalysis, nanobarcode, quantum dots etc. It is a rapidly growing industry currently worth billions of U.S. dollars, with many potential benefits to society. Nanotechnology has the environmental applications to protect climate change and environmental pollution as nanocatalyst, light weight nanocomposite materials, nanocoatings, improved renewables and nanosensors.

This fast growing technology is already having a significant commercial impact, which will certainly increase in the future. Because of their widespread application, the commercial nanotechnology industry is predicted to increase significantly to more $3 trillion in few years. While nanotechnologies offer many opportunities for innovation, the use of nanomaterials in food, agriculture and environment has also raised a number of safety, environmental, ethical, policy and regulatory issues.

What are Nanoparticles?

Nanoparticles (NPs) are clusters of atoms in the size range of 1–100 nm. "Nano" is a Greek word synonymous to dwarf meaning extremely small. The word "nano" is used to indicate one billionth of a meter or 10^{-9}. The ideas and concepts behind nanoscience and nanotechnology started with a talk entitled "There's Plenty of Room at the Bottom" by physicist Richard Feynman at an American Physical Society meeting at the California Institute of Technology on December 29, 1959, long before the term nanotechnology was used. In his talk, Feynman described a process in which scientists would be able to manipulate and control individual atoms and molecules. Over a decade later, the term Nanotechnology was coined by Professor Norio Taniguchi of Tokyo Science University, Japan in the year 1974 to describe precision manufacturing of materials at the nanometer level. Bio-nanotechnology has emerged up as integration between biotechnology and nanotechnology for developing biosynthetic and environmental-friendly technology for synthesis of nanomaterials. The use of nanoparticles is gaining impetus in the present century as they possess defined chemical, optical and mechanical properties. The metallic nanoparticles are most promising as they show good antibacterial properties due to their large surface area to volume ratio, which is coming up as the current interest in the researchers due to the growing microbial resistance against metal ions, antibiotics and the development of resistant strains. Different types of nanomaterial like copper, zinc, titanium, magnesium, gold, alginate and silver have come up but silver nanoparticles have proved to be most effective as it has good antimicrobial efficacy against bacteria, viruses and other eukaryotic micro-organisms

Nanotechnology in Agriculture

Attempts to apply nanotechnology in agriculture began with the growing realization that conventional farming technologies would neither be able to increase productivity any further nor restore ecosystems damaged by existing technologies back to their pristine state; in particular because the long-term effects of farming with "miracle seeds", in conjunction with irrigation, fertilizers, and pesticides, have been questioned both at the scientific and policy levels, and must be gradually phased out. Nanotechnology in agriculture has gained momentum in the last decade with an abundance of public funding, but the pace of development is modest, even though many disciplines come under the umbrella of agriculture. This could be attributed to: a unique nature of farm production, which functions as an open system whereby energy and matter are exchanged freely; the scale of demand of input materials always being gigantic in contrast with industrial nanoproducts; an absence of control over the input nanomaterials in contrast with industrial nanoproducts (eg, the cell phone) and because their fate has to be conceived on the geosphere (pedosphere)-biosphere-hydrosphere-atmosphere continuum; the time lag of emerging technologies reaching the farmers' field, especially given that many emerging economies are unwilling to spend on innovation; and the lack of foresight resulting from agricultural education not having attracted a sufficient number of brilliant minds the world over, while personnel from kindred disciplines might lack an understanding of agricultural production systems. If these issues are taken care of, nanotechnologic intervention in farming has bright prospects for improving the efficiency of nutrient use through nanoformulations of fertilizers, breaking yield barriers through bionanotechnology, surveillance and control of pests and diseases, understanding mechanisms of host-parasite interactions at the molecular level, development of new-generation pesticides and their carriers, preservation and packaging of food and food additives, strengthening of natural fibers, removal of contaminants from soil and water, improving the shelf-life of vegetables and flowers, clay-based nanoresources for precision water management, reclamation of salt-affected soils, and stabilization of erosion-prone surfaces, to name a few.

Nanotechnology and Animal Health

Nanotechnology, as a new enabling technology, has the potential to revolutionise agriculture and food systems in the United States of America and throughout the world. Examples of potential applications of nanotechnology in the science and engineering of agriculture and food systems include disease treatment delivery systems, new tools for molecular and cellular biology, the security of agricultural and food systems, new materials for pathogen detection, and protection of the environment. Existing research has clearly demonstrated the feasibility of introducing nanoshells and nanotubes into animal systems to seek out and destroy targeted cells. Nanoparticles smaller than one micron have been used to deliver drugs and genes into cells. Thus, some building blocks do exist in isolation and are expected to be integrated into systems over the next 10 to 15 years. It is reasonable to presume over the next couple of decades that nanobiotechnology industries and unique developments will revolutionise animal health and medicine.

Nanotechnology in BioMedical Applications

Nanotechnology is rapidly expanding into the biomedical field. At the Institute for Lasers, Photonics and Biophotonics, research in nanotechnology has focused on the development of surface functionalized nanoparticles for diagnostics and targetted therapy. This technology provides a platform for the development of new imaging, diagnostic and therapeutic modalities. For bioimaging and diagnostics, nanoparticles are fabricated containing rare-earth ions which exhibit two-photon, anti-stokes luminescence by frequency up-converting infrared to visible light. We have successfully prepared the phosphor containing nanoparticles having a size ~25 nm with a silica shell around it which helps in aqueous dispersabilty, inhibits water quenching with nanophosphors and allows functionalization for covalent binding of bioprobes for targeting. The coupling of specific peptides, proteins or nucleic acid sequences to the silica shell will allow for the selective detection of biological entities and will have applications in bioimaging, flow cytometry, ELISA and DNA/RNA hybridization systems. Therapeutically, this nanoparticle platform was used to develop "nanoclinics" which can selectively target specific cancer cells. Our prototypic nanoclinic utilizes magnetocytolysis to effect distruction of luteinizing hormone-releasing hormone (LH-RH) receptor positive cancers. Bioadhesive nanoparticles can also provide a novel mechanism for controlled drug delivery. Presently only 1-3 % of the topically administered Brimonidine penetrates the cornea and reaches intraocular tissues. A nanoparticle of the bioadhesive polymer, polyacrylic acid (PAA), incorporating the ocular drug, Brimonidine, has been prepared. This formulation would prolong drug contact in the precorneal area resulting in reduced drug cost to the patient.

Nanotechnology to Combat Climate Change

Recently, German Chancellor, Angela Merkel, told lawmakers that tackling climate change will be one of the central tasks of the upcoming Hamburg G20 summit of the world's largest economies, following the U.S.'s withdrawal from the Paris climate pact. Here we look at the ways that nanotechnology can help to combat and possibly stop climate change:

1. Lightweight nano-composite materials: Any effort to reduce emissions in vehicles by reducing their weight , in turn, decreasing fuel consumption can have an immediate and significant global impact. It is estimated that a 10% reduction in weight of the vehicle corresponds to a 10% reduction in fuel consumption, leading to a proportionate fall in emissions. In recognition of the above, there is growing interest worldwide in exploring means of achieving weight reduction in automobiles through use of novel materials. For example, use of lighter, stronger, and stiffer nano-composite materials is considered to have the potential to significantly reduce vehicle weight.

2. Nano-coatings: Nanotechnology coatings are a good short-term way of reducing emissions and and maximizing clean energy production. For example, nano-coatings can be applied to aircraft, which can make aircraft's smoother, reducing drag and also protect the materials from the special conditions of the environment where they are used (instead of the conventional bulk metals such as steel). Since the amount of CO2 emitted by an aircraft engine is directly related to the amount of fuel burned, CO2 can be reduced by making the airplane lighter. Hydrophobic nano-coatings can also improve the energy produced from solar panels for example.

***3. Nanocatalysts*:** Nanotechnology is already applied to improve fuel efficiency by incorporation of nanocatalysts. Enercat, a third generation nanocatalyst developed by Energenics, uses the oxygen storing cerium oxide nanoparticles to promote complete fuel combustion, which helps in reducing fuel consumption. Recently, the company has demonstrated fuel savings of 8%–10% on a mixed fleet of diesel vehicles in Italy. Reducing friction and improving wear resistance in engine and drive train components is of vital importance in the automotive sector. Based on the estimates made by a Swedish company Applied Nano Surfaces, reducing friction can lower the fuel consumption by about 2% and result in cutting down CO2 emissions by 500 million tons per year from trucks and other heavy vehicles in Sweden alone.

4. Nano-structured Materials: Thanks to nanomaterials like silica, many tires will in the future be capable of attaining the best energy rating, the green category. Cars equipped with category A tires consume approximately 7.5% less fuel than those with tires of the minimum standard (category G). Residential and commercial buildings contribute to 11% of total greenhouse gas emissions. Space heating and cooling of residential buildings account for 40% of the total residential energy use. Nanostructured materials, such as aerogels, have the potential to greatly reduce heat transfer through building elements and assist in reducing heating loads placed on air-conditioning/heating systems. Aerogel is a nanoporous super-insulating material with extremely low density; silica aerogel is the lightest solid material known with excellent thermal insulating properties, high temperature stability, very low dielectric constant and high surface area.

5. Improved Renewables: Nanotechnology may accelerate the technology behind renewables in various ways:experts are discovering means to apply nanotechnology to photovoltaics, which would produce solar panels with double or triple the output by 2020; wind turbines stand to be improved from high-performance nano-materials like graphene, a nano-engineered one-atom thick layer of mineral graphite that is 100 times stronger than steel. Nanotechnology will enable light and stiff wind blades that spin at lower wind speeds than regular blades;nanotechnology could play a major role in the next generation of batteries. For example, coating the surface of an electrode with nanoparticles increases the surface area, thereby allowing more current to flow between the electrode and the chemicals inside the battery.

***6. Nanoclay materials*:** The emphasis for future work is on advanced clay-based nanomaterials for use in new approaches to sustainable energy, green environment, and human health. One such application of Sodium bentonite nanoclay to reduce temperature, absorption of toxic dissolved gases and electrolytes from water environment is a promising concept to manage confined environment.

***7. Batteries*:** Such techniques could increase the efficiency of electric and hybrid vehicles by significantly reducing the weight of the batteries. Nanotechnology is positioned to create significant change across several domains, especially in energy where it may bring large and possibly sudden performance gains to renewable sources and Smart Grids. Nanotech enhancements may also increase battery power by orders of magnitude, allowing intermittent sources such as solar and wind to provide a larger share of overall electricity supply without sacrificing stability. Moreover, superior batteries would complement renewables by storing energy economically, thus offsetting the whole issue of intermittent generation. In a somewhat more distant future, we may see electricity systems apply nanotechnology in transmission lines. Research indicates that it is possible to develop electrical wires using carbon nanotubes that can carry higher loads and transmit without power losses even over hundreds of kilometers. The implications are significant, as it would increase the efficiency of generating power where the source is easiest to harness. Semiconductor devices, transistors, and sensors will benefit from nanotechnology especially in size and speed.

8. Nanotech sensors: Sensors could be used for the Smart Grid to detect issues ahead of time, i.e., to measure degrading of underground cables or to bring down the price of chemical sensors already available for transformers. Nanotechnology will likely become indispensable for the Smart Grid to fully evolve in the near future. Energy efficiency is a way of managing and restraining the growth of energy consumption. It is one of the easiest and most cost effective ways to combat climate change, improve the competitiveness of businesses, and reduce energy costs for consumers.

ICAR-CIFA has undertaken research on nanotechnology since last one decade. The institute has collaborated with seven Universities, nine national institutes and two IITs in production and research application of nanomaterial, besides is in the process of commercialising two Nano-based formulations (Nanox: Antibiofouling and Wound Healing Nano formulations for wound healing and Zinc-Selenium Nanoplus for Fish Feed to improve growth and disease resistance) through Agrinnovate, A company under Department of agricultural Research and Education, Ministry of agriculture, Govt of India. Process of synthesis of nanoparticles such metal and metal oxides, inorganic and polymeric materials have been developed. Their uses include treatment and remedial of water qualities; fish health management including disease diagnosis, wound healing and immune modulations; control of bacterial biofilm, fungal infections and algal blooms; enhancing the keeping quality of fish feed and fish products; enrichment of micronutrient content of fish flesh; skin coloration and quality improvement of aquarium fishes; hatchery applications for enhancing hatching and survivality of fish fry. There are more opportunities for its use in fishery, aquaculture and allied sectors in future, besides a lot of career prospects for graduates and technocrats.

References

Adesina, A.A. 2004. Industrial exploitation of photocatalysis: Progress, perspectives and prospects. Catal. Surv. Asia, 8(4): 265-273.

Alfvén, T., Järup, L. and Elinder, C.G. 2002. Cadmium and lead in blood in relation to low bone mineral density and tubular proteinuria. Environ Health Persp., 110: 699-702.

Alivisatos, A.P. 2001. Less is more in medicine. Scientific American, 9: 59-65.

Alvarez-Ayuso, E., Garcia-Sanchez, A. and Querol, X. 2003. Purification of metal electroplating waste waters using zeolites. Water Res., 37: 4855-4862.

Aorkas, M., Tsiourvas, D. and Paleos, C.M. 2003. Functional dendrimeric "nanosponges" for the removal of polycyclic aromatic hydrocarbons from water. Chem. Mater, 15(14):2844-2847.

Asahi, R., Morikawa, T., Ohwaki, T., Aoki, K. and Taga, Y. 2001. Visible-light photocatalysis in nitrogen-doped titanium oxides. Science, 293(5528): 269-271.

Auffan, M., Rose, J., Wiesner, M.R. and Bottero, J.Y. 2009. Chemical stability of metallic nanoparticles: A parameter controlling their potential cellular toxicity *in vitro.* Environmental Pollution, 157: 1127-1133.

Bae, E. and Choi, W. 2003. Highly enhanced photoreductive degradation of perchlorinated compounds on dye-sensitized metal/TiO2 under visible light. Environ. Sci. Technol., 37(1): 147-152.

Baesman, S.M., Bullen, T.D., Dewald, J., Zhang, D., Curran, S., Islam, F.S., Beveridge, T.J. and Oremland, R.S. 2007. Formation of Tellurium Nanocrystals during Anaerobic Growth of Bacteria that Use Te Oxyanions as Respiratory Electron Acceptors. Applied and Environmental Microbiology, 73(7): 2135-2143.

Bruchez M, Moronne M, Gin P, Weiss S, Alivisatos AP. 1998. Semiconductor nanocrystals as fluorescent biological labels. Science. ;281:2013–2016. doi: 10.1126/Science.281.5385.2013.

Chan WCW, Nie SM. 1998. Quantum dot bioconjugates for ultrasensitive nonisotopic detection. Science.;281:2016–2018. doi: 10.1126/science.281.5385.2016.

Feynman R. 1991. There's plenty of room at the bottom. Science. ;254:1300–1301.

Keren K, Berman RS, Buchstab E, Sivan U, Braun E. DNA-templated carbon nanotube field-effect transistor. Science. 2003;302:1380–1382. doi: 10.1126/science.1091022.

Mah C, Zolotukhin I, Fraites TJ, Dobson J, Batich C, Byrne BJ. 2000. Microsphere-mediated delivery of recombinant AAV vectors in vitro and in vivo. Mol Therapy. ;1:S239. doi: 10.1006/mthe.2000.0174.

Mazzola L. 2003. Commercializing nanotechnology. Nature Biotechnology. ;21:1137–1143. doi: 10.1038/nbt1003-1137.

Murray CB, Kagan CR, Bawendi MG. 2000. Synthesis and characterisation of monodisperse nanocrystals and close-packed nanocrystal assemblies. Annu Rev Mater Sci. ;30:545–610. doi: 10.1146/annurev.matsci.30.1.545.

Panatarotto D, Prtidos CD, Hoebeke J, Brown F, Kramer E, Briand JP, Muller S, Prato M, Bianco A. 2003. Immunization with peptide-functionalized carbon nanotubes enhances virus-specific neutralizing antibody responses. Chemistry&Biology. ;10:961–966.

Pankhurst QA, Connolly J, Jones SK, Dobson J. Applications of magnetic nanoparticles in biomedicine. J Phys D: Appl Phys. 2003; 36:R167–R181. doi: 10.1088/0022-3727/36/13/201.

Parak WJ, Gerion D, Pellegrino T, Zanchet D, Micheel C, Williams CS, Boudreau R, Le Gros MA, Larabell CA, Alivisatos AP. 2003. Biological applications of colloidal nanocrystals. Nanotechnology. ;14:R15–R27. doi: 10.1088/0957-4484/14/7/201.

Paull R, Wolfe J, Hebert P, Sinkula M. 2003. Investing in nanotechnology. Nature Biotechnology. 21:1134–1147. doi: 10.1038/nbt1003-1144.

Taton TA. Nanostructures as tailored biological probes. Trends Biotechnol. 2002;20:277–279. doi: 10.1016/S0167-7799(02)01973-X.

Wang S, Mamedova N, Kotov NA, Chen W, Studer J. 2002. Antigen/antibody immunocomplex from CdTe nanoparticle bioconjugates. Nano Letters. ;2:817–822. doi: 10.1021/nl0255193.

Whitesides GM. The 'right' size in Nanobiotechnology. Nature Biotechnology. 2003;21:1161–1165. doi: 10.1038/nbt872.

Yan H, Park SH, Finkelstein G, Reif JH, LaBean TH. 2003. DNA-templated self-assembly of protein arrays and highly conductive nanowires. Science. ;301:1882–1884. doi: 10.1126/science.1089389.

28

Proteomics as a Potential Tool to Study the Effects of Climate Change in Aquatic Organisms

Mohan R Badhe and J Mohanty

Introduction

Aquatic ecosystems are critical components of the global environment and provide a variety of resources for human population, including food, water for drinking and irrigation, recreational opportunities, and habitat for economically important fisheries. Fish is the main source of animal protein for 3 billion people worldwide, providing a valuable protein complement to the food platter worldwide. Fish is also an important source of essential vitamins and fatty acids. Furthermore, it provides an important source of cash income for many poor households and holds a great potential as a source of foreign exchange. However, aquatic ecosystems are increasingly being threatened directly and indirectly by climate change. Global warming with consequent climate variability and extreme weather conditions has a great impact on the aquatic ecosystems resulting in increased temperature variation, ocean acidification and freshwater salinization etc. As per report of IPCC 2018, global warming of 1.5 °C above pre-industrial levels has been declared (IPCC, 2018) and expected to rise further in future. Thus, variations in temperature, salinity, acidity and dissolved oxygen along with the effects of other anthropogenic activities may exert strong influences on aquatic ecosystems and the aquatic animls will be facing these climate-related stressors in near future. Such environmental stressors is projected to alter fundamental ecological processes, geographic distribution of aquatic species, reproductive and immunological performance of species etc.

Any stress response in fish involves principal messengers of the brain-sympathetic-chromaffin cell axis and the brain-pituitary-interrenal axis. The responses associated with these axes involve stimulation of oxygen uptake and transfer, mobilization of energy substrates and mainly suppressive effects on immune functions. There are three stages of stress response in fishes *viz.*, primary, secondary and tertiary. Primary responses involve the initial neuroendocrine responses, release of catecholamines (Randall & Perry 1992; Reid *et al.* 1998), and the stimulation of the hypothalamic-pituitary-interrenal axis resulting in the release of corticosteroid hormones into circulation (Donaldson 1981; Bonga 1997). Secondary responses cause variations in plasma levels, ion concentration in tissues and associated metabolic changes along with elevated levels of heat shock proteins. This results in altered metabolism, respiration, hydromineral balance, immune function and cellular responses (Pickering 1981; Iwama *et al.* 1998). Furthermore, tertiary responses cause growth variations, overall resistance to disease, behavior, and ultimately survival (Wedemeyer & McLeay 1981; Wedemeyer *et al.* 1990).

Proteomics

Proteomics is a recent field in molecular biology that involves the study of all proteins present in a cell or tissue or whole organism. It includes protein identification and quantification, their structure and function as well as their interactions with other biomolecules. In eukaryotes, the proteome is bigger than the genome, mainly due to alternative splicing of genes and various post-translational modifications (PTMs) like acetylation, phosphorylation, and glycosylation. Proteome analysis is increasingly gaining importance as protein components are directly related to cellular functions. It provides knowledge about changes occurring at the level of the proteins including the processes of synthesis, posttranslational modifications, and degradation; in response to various stimuli. In recent years, proteomic techniques have started to become very useful tool in a variety of model systems including fisheries research. Across a number of disciplines within fisheries science, there is now a growing interest in understanding the cellular mechanisms of organism-level responses to changes in the physical and chemical environment particularly of temperature, osmolality and oxygen concentration. This chapter provides a brief overview of the basic approach and the methodological tools available to study different aspects of the proteome. Furthermore, how a proteomics approach can be a powerful tool for hypothesis-driven investigations on a number of exciting fronts in basic and applied response to climate change has also been explained.

Proteomics strategies for the identification and analysis of proteins

***Gel based proteomics*:** A proteomic study involves a large variety of different protein molecules to be studied simultaneously and hence requires high-throughput methodologies for analysis. Two-dimensional gel electrophoresis (2-DE) is one of the most widely used techniques to obtain data on protein expression levels under various conditions. In 2-DE the protein samples are first separated by isoelectrofocusing in one dimension followed by SDS-PAGE in 2nd dimension resulting in a pattern of spots. Thus all proteins are separated according to their two independent properties first by isoelectric points (pI) and second by molecular size. The development of immobiline pH gradient (IPG) strips provides a precise and stable pH gradient for the first dimension IEF that helps in generation of a more reproducible 2D maps. The 2D gels are then stained and the protein spots are analyzed by commercial 2D softwares. Each spot in a 2D gel represents one polypeptide and the intensity of spot indicates the quantity of the particular protein present in the sample. The spots of interest can be excised from the gel, subjected to in-gel digestion with trypsin and the resulting fragments are exposed to mass spectrometry (MS) analysis for identification of the protein. Several thousands of proteins have been successfully resolved on a single gel using this technique. Thus, it enables identification of differentially expressed proteins between types of specimens run in separate gels. However, low sensitivity, narrow linear dynamic range of detection by staining methods and poor reproducibility of 2-DE have made the high-throughput quantitation of proteins considerably challenging.

The problem of comparison of images across 2-DE gels has been overcome by difference gel electrophoresis (DIGE) where cyanine dyes, Cy3 and Cy5, are used for differential labeling of the same protein from different samples. These dyes are designed to be charge- and size-matched so that charge state of the labeled proteins are maintained and it allows precise co-migration of differentially labeled protein species within the same 2-D gel. However, due to spectrally distinct fluorophores of Cy3 and Cy5 it allows detection of the signal from one sample without appreciable contribution from the other. Cy2 dye is used to label an internal standard, which consists of a pooled sample comprising of equal amounts of control and treatment samples. Relative protein abundance of samples can be readily obtained by simple comparison of their fluorescence intensities using a fluorescence gel scanner. Scanning at different wavelengths, characteristic for each dye, permits to acquire three images from only one gel. Even small differences in expression levels can be determined by comparing the ratio obtained from one fluorescent-labeled sample directly with other using different wavelengths for image analysis. DIGE allows the analysis of two or more protein samples simultaneously on a single two-dimensional gel, which is not possible with other 2-DE techniques.

In addition to fluorescent labels, radioactive isotopes have also been employed for labeling procedures and subsequent protein detection and identification in gel-based proteomics. Isotopic labeling techniques offer a less destructive means for the protein of interest without compromising on the detection sensitivity. Coupling of 2-DE with mass spectrometry (MS) in recent years has facilitated the identification of various protein complexes. In spite of the significant advancements in gel-based methods, detection of low abundance proteins as well as hydrophobic proteins has proved challenging, due to which many gel-free proteomic approaches have evolved.

***Gel free proteomics*:** Protein separation using liquid chromatography and electrophoresis has shown tremendous progress to enable resolution and identification of digested peptides of a proteome. This is commonly referred as liquid chromatography-mass spectrometry (LC-MS) methods. By LC-MS methods, the proteins are first digested enzymatically to produce peptide fragments, followed by liquid chromatography and tandem mass spectrometry to identify protein sequences. Procedures applicable for such complex resolution of several thousands of proteins/peptides are mainly high-performance liquid chromatography (HPLC) and capillary electrophoresis (CE). Ion exchange, strong cation exchange (SCX), reverse phase (RP), size exclusion and hydrophobic interaction are the various modes of HPLC available while capillary zone, isoelectric focusing (IEF), affinity and micellar electrophoresis are the different types of CE commonly used. However, it is quite well known that no single chromatographic or electrophoretic separation strategy is by itself sufficient to resolve a complex mixture of peptides. A combination of two of more separation procedures, which must be orthogonal and compatible with each other, is therefore employed to improve the overall resolution of complex proteome digests. One such successful combination method developed and widely used is the Multidimensional Protein Identification Technology (MuDPIT), which makes use of strong cation exchange chromatography (SCX) in one dimension and reverse phase HPLC in the other dimension. Combinations such as RP-HPLC with capillary zone electrophoresis (CZE), affinity and RP chromatography and several others have been developed, each with their own advantages and limitations. The ultimate objective of these separation protocols is to reduce the sample complexity in order to facilitate efficient and accurate identification by mass spectrometry.

***Mass spectrometry*:** Mass spectrometry is a technique that allows the detection of compounds by separating ions by their unique mass (mass-to-charge ratio) using a mass spectrometer. There are many types of mass spectrometers that can be used for proteomic studies, and each accomplishes the task of peptide identification in a slightly different way. However, the basic process of identifying a protein using a mass spectrometer is consistent between the various types. After initial protein digestion typically with trypsin, peptides must be ionized to enter the mass spectrometer. Peptides are then detected, isolated, and finally, fragmented and sequenced by the mass spectrometer.

Current mass spectrometers can detect and identify peptides in the femtomole (10^{-15}) to attomole (10^{-18}) range.

Ionization of peptides is the first step in mass spectrometry of proteomes. The two frequently used ionization methods are electrospray ionization (ESI) and matrix-assisted laser desorption ionization (MALDI). One advantage of ESI is that this method of ionization allows for the direct linkage between liquid chromatography and mass spectrometry because of the volatility of the HPLC solutions. Charged gas phase peptides are generated by ESI when the acidic HPLC solution containing peptides is sprayed from a tip, and the solution evaporates. The MALDI requires mixing of the peptide with a UV-absorbing molecule and the formation of crystals. When a laser strikes the crystalline structure, the results are the sublimation of the matrix and ionization and release of the associated peptides. The peptides are then analyzed by the mass spectrometer and the peptide mass determined. The peptide's mass is typically expressed as a ratio of mass divided by the charge of the peptide (m/z).

There are two basic mass spectrometry methods for the identification of proteins. The first called peptide fingerprinting is often associated with 2D-PAGE protein separation scheme. Individual spots from the 2D-PAGE gel are isolated and the proteins digested with typsin. These proteins are typically analyzed with a MALDI-TOF (time of flight) mass spectrometer. The mass spectrometer will record all the peptide m/z detected in the gel spot. Identification of a protein is based on the measurement of multiple peptides that can come from that protein. For example, after a mass spectrometer has determined the m/z for the peptides from a gel spot, this information will be matched to a protein database. A successful protein match will be based on the number of peptides matched to the protein and the accuracy of the matches. The second method to identify proteins involves use of tandem mass spectrometers that allow sequencing of individual peptides.

Quantitative proteomics: Besides the global profiling of the proteins present in a system at a given time, quantification of the proteins is necessary to understand the level of protein expression and how their abundances can change in response to an altered state. Most quantitative proteomic analyses entail the isotopic labeling of proteins or peptides in the experimental groups, which can then be differentiated by mass spectrometry. Some commonly used quantitative techniques are detailed below.

Isotope-Coded Affinity Tags (ICAT): ICAT is a quantitative proteomics method based on the use of a new class of chemical reagents termed isotope-coded affinity tags and tandem mass spectrometry. These chemical probes consist of three general elements i.e. defined amino acid side chain, an isotopically coded linker, and a tag for the affinity isolation of labelled proteins/peptides. In this method, the relative quantification is highly accurate because it is based on stable isotope dilution techniques. The ICAT approach should provide a widely applicable means to compare quantitatively global protein expression in cells under different conditions or at different stages of the biological sample.

Isobaric tags for relative and aqsolute quantitation (iTRAQ): iTRAQ is multiplex protein labeling technique for protein quantification based on tandem mass spectrometry. This technique relies on labeling the protein with isobaric tags (4-plex or 8-plex) for relative and absolute quantitation. Based on the covalent labelling of the N-terminus and side chain amines of peptides with tags of varying mass, 4-plex and 8-plex reagents can be used to label all peptides from different samples. Since the tags have the same complete mass, each peak detected in MS represents a single peptide from the combined samples. MS/MS of each peptide releases the reporter allowing simultaneous quantitation and identification of the peptide. Furthermore, iTRAQ labeling does not rely on the presence of cysteine; therefore, the main advantage of the method is the ability to visualize multiple peptides from a single protein and hence achieve much stronger intra-experimental statistical viability, further reducing analysis time.

Stable Isotope Labeling with Amino acids in Cell Culture (SILAC): SILAC is an MS-based approach for quantitative proteomics that depends on metabolic labeling of whole cellular proteome. It is similar to those described above, except that cells subjected to different biological conditions are grown in culture in the presence of an essential amino acid with a stable isotopic nucleus. Therefore, one sample is incubated with an unlabelled amino acid and the test sample incubated with a deuterated form. Because the amino acid is essential, the organism requires it for survival, and through several replication cycles all of that particular amino acid will be present in the cells' proteins in either unlabelled (control) or deuterated (test) form, allowing true quantitation.

***Bioinformatics in proteomics*:** Bioinformatics plays a major role in the study of proteomics enhancing our understanding of protein sequence, structure, function, regulation, expression and modification. Besides, it helps in management of large amount of data generated by high-throughput technologies that require fast and profound analyses, and data mining as well.

Several protein databases generated provide information related to protein characterization, domain and families, sequences, biophysical and catalytic properties, function and applications, identification and quantification, annotations, expression and profiling, 2D/3D structure and homology models, protein-protein and protein-ligand interactions, pathways and networks, post translational modifications, structure predictions etc. These databases help in extracting relevant information from large data sets and understanding the statistical significance of protein identification results. The development of protein databases is important for discovery of signature candidates for biomarkers, drug targets, therapeutics, and mapping protein interaction networks.

Besides generation of protein databases, several bioinformatics tools are also available for MS data analysis and identification of proteins. Sequence databases are searched to identify proteins, which match the peptide mass fingerprints generated by MALDI-MS. The growth of databases along with genome sequencing information adds to the great accuracy of the protein identification. Once the complete genome for an organism is available, all the fingerprint patterns can be predicted and detected accurately.

Proteomics in response to environmental stress

Any stimulus that impairs the normal performance or affects the homeostasis of an animal can be called as stressors. The exposure to the stressors can be acute or chronic. The potential stressors to fish can be grouped as environmental, physical and biological. Environmental stressors mainly include adverse physical or chemical conditions of water. Extreme condition or change in water qualities such as dissolved oxygen, ammonia, hardness, pH and temperature can stress the fish. Fish respond to these stresses in highly ordered manner regulated by the neural, endocrine, and immune systems. The primary function of these responses is to compensate fish biological systems, arrange the metabolism to afford the energy required by the fish and maintain homeostasis. A broad range of metabolic processes and pathways are involved in this stress response and proteomics provides an excellent platform for studying such metabolic changes. Thus, the fitness of an organism to stressors can be assessed through monitoring their proteome for the molecular traits that are directly related to phenotypical performance. The following studies elucidate the changes occuring in the proteome framework of aquatic organisms mostly in fish, experimentally exposed to commonly occurring environmental stressors such as temperature, oxygen and salinity.

Thermal Stress: It is well established that temperature beyond optimum limits of a particular species adversely affect the health of aquatic animal due to metabolic stress and increases oxygen demand and susceptibility to various harmful pathogens. Hence, thermal tolerance studies have generated significant interest in researchers worldover to understand the impact of global warming on animals, including fish.

Nuez-Ortín *et al.* (2018) assessed the variation in the liver proteome of Atlantic salmon with control (15 °C) and elevated (21 °C) temperature treatments for 43 days using NanoLC-MS and found 276 proteins to be differentially expressed. As identified by Ingenuity Pathway Analysis (IPA), transcription and translation mechanisms, protein degradation via proteasome, and cytoskeletal components were down-regulated at elevated temperature. In contrast, an up-regulated response was identified for NRF2 (nuclear factor erythroid 2-related factor 2)-mediated oxidative stress, endoplasmic reticulum stress, and amino acid degradation.

The liver proteome of murrel *Channa striatus* were analysed after the exposure to heat stress at 36 °C for 4 days by using gel-based proteomics, i.e. 2DE followed by MALDI-TOF/TOF-MS (Mahanty *et al.* 2016). Increased abundance of two sets of proteins, the antioxidative enzymes superoxide dismutase (SOD), ferritin, cellular retinol binding protein (CRBP), glutathione-S-transferase (GST), and the chaperone HSP60 and protein disulfide isomerase were observed. In skeletal muscle tissue of another freshwater fish, common carp *Cyprinus carpio*, cold acclimation (10 °C versus 30 °C) showed an increased accumulation of fragments of creatine kinase (McLean *et al.* 2007).

Proteomic changes in liver tissue of sea bream *Sparus aurata* in response to cold-stress (8 °C) in comparison to normal (20 °C) have been analysed, which showed many proteins to be down-regulated including actin, the most abundant protein in the proteome; enzymes of amino acid metabolism; and enzymes with antioxidant capacity, such as betaine-homocysteine-methyl transferase, glutathione-S-transferase and catalase (Ibarz *et al.* 2010). They also showed upregulation of proteases, proteasome activator protein and trypsinogen-like protein indicating an increase in proteolysis. Besides, increases in elongation factor-1alpha, the GAPDH oxidative form, tubulin and Raf-kinase inhibitor protein indicative of oxidative stress and the induction of apoptosis were also observed. These data indicated that cold exposure induced oxidative damage in hepatocytes. In another study in *S. aurata*, exposed to different temperature regimes, (control 18 °C, nursery ground temperature 24 °C and heat wave 30 °C); fish subjected to 30 °C showed enhanced glycolytic potential and up-regulated proteins related to gene expression, cellular stress response, and homeostasis (mostly cytoskeletal dynamics, acid-base balance, chaperoning) (Madeira *et al.* 2017).

Thus, acute and chronic temperature stresses induce wide-ranging proteomic changes in protein synthesis and degradation that are accompanied by increasing levels of chaperones. Oxidative stress levels vary greatly in parallel with major changes in energy metabolism.

Hypoxic stress: This condition of low dissolved oxygen (DO) is known as hypoxia while a condition with negligible amount of oxygen is known as anoxia. Generally hypoxia can be defined as dissolved oxygen less than 2.8 mg O_2/L (equivalent to 2 mL O_2/L or 91.4 mM) (Diaz & Rosenberg, 1995). The condition can be observed in a variety of marine, estuarine, and freshwater habitats. Its occurrence is primarily due to excessive anthropogenic input of nutrients and organic matters into water bodies with poor circulation.

Hypoxia has already led to major changes in fish species composition, alteration of food webs and community structure, decrease in species richness and diversity, population declines and extinction of sensitive species in both marine and freshwater systems in many parts of the world. In general, freshwater systems are more prone to hypoxia and anoxia, with a long history of occurrence. Lakes and coastal areas with seasonal stratification tend to be highly sensitive to anthropogenic nutrient enrichment.

Hypoxia can induce reduced growth, alter behaviors and changes in food preferences thereby reducing their abundance and diversity. It also impairs reproductive success by affecting a number of key reproductive processes, including gametogenesis, sperm and egg count and quality, reproductive behaviors, fertilization success, hatching and ultimately, larval fitness of juveniles. These damages may be recovered through disrupting the various hormones and enzymes regulating these key reproductive processes.

In the aquatic animals including finfish and shellfish, hypoxic stress lowers levels of a number of proteins involved in energy metabolism, antioxidant processes, chaperoning, and cytoskeleton, while increasing abundances of proteins involved in the immune response. Limited proteomic changes were observed in zebrafish skeletal muscle in response to hypoxic conditions (oxygen 16 torr) for 48 hours (Bosworth *et al.* 2005). Whereas, glycolytic suppression has been shown to occur during hypoxia in muscle tissue of kuruma prawn (*Marsupenaeus japonicus*) through the downregulation of fructose-bisphosphate aldolase, a glycolytic protein (Abe *et al.* 2007). Chen *et al.* (2013) used 2D-DIGE to study protein expression pattern in white skeletal muscle of zebrafish after 48 h exposure to hypoxia (Po_2=19kPa) and reported higher expression of several glycolytic enzymes in hypoxia than in normoxia, whereas enzymes associated with mitochondrial ATP synthesis were lower during hypoxia. Among the more highly up-regulated proteins during hypoxia were two variants of hemoglobin α subunit. The results were in agreement with hypoxic response that enhances anaerobic metabolism and O_2 transport to tissues, with concomitant suppression of mitochondrial metabolism.

Some fish species are highly tolerant to hypoxia, due to adaptation protecting the brain from the metabolic consequences of hypoxic conditions (Hochachka & Lutz, 2001). Reductions in the levels of proteins functioning in glycolysis, neuronal apoptosis (voltage-dependent anion channel), and neural degeneration (dihydropyrimidinase-like protein 3 and vesicle amine transport protein 1) were observed in proteomic analysis of brain tissue of crucian carp (*Carassius carasssius*). These types of responses suggest that the brain becomes more protected against cellular injury (Smith *et al.* 2009). Proteomic analysis of shark *Hemiscyllium ocellatum* during hypoxic condition showed decreases in proteins like glutaminase involved in glutamate production and, release and changes of proteins associated with the regulation of internal calcium stores in the cerebellum (Dowd *et al.* 2010). On the other hand, in species that do not tolerate hypoxic conditions, the proteins representing similar cellular functions tend to increase rather than reduce. However, the hypoxia-tolerant medaka (*Oryzias latipes*) showed an elevation in glycolytic enzyme levels involved in Krebs cycle such as aldolase in brain tissue in response to hypoxia (Oehlers *et al.* 2007). Besides, increasing levels of hemoglobin and carbonic anhydrase suggested enhanced blood flow and vascularization, which might improve the delivery of existing levels of blood oxygen.

Proteomic studies on hypoxic stress are providing evidence for enzymatic changes involved in energy metabolism in organisms that are hypoxia-tolerant and such type of changes occur in a number of tissues simultaneously. Such studies might help in elucidating the regulatory mechanisms of the signaling events that activate the cellular response to hypoxia.

***Salinity stress*:**Another key stressor involved in climate change scenario is salinity change. Freshwater salinization is caused by primary as well as secondary salinization. The natural drivers of salinization includes rainfall, rock weathering, aerosol deposits, seawater intrusion etc., which contribute to primary salinization while anthropogenic factors such as construction, agriculture runoff, resource extraction, saline groundwaters coming to surface comes under secondary salinization. Organisms may encounter changing salinity (osmotic) conditions near shore due to proximity to estuaries and major rivers. Fish are least tolerant to salinity in their juvenile stages. Higher salinity levels put the cells of many organisms under osmotic stress resulting in loss of pH regulation or Na poisoning.

Ky *et al.* (2007) analysed the gill and intestine epithelial proteomes of European sea bass (*Dicentrarchus labrax*) reared under seawater and freshwater conditions for a period of 3 months. It showed an overabundance of cytoskeletal proteins and aromatase cytochrome P450 in gills when fish were reared in seawater, whereas prolactin receptor and major histocompatability complex class II β-antigen were higher when fish were reared in freshwater. In intestinal cells under freshwater conditions, the Iroquois homeobox protein Ziro5 was upregulated. In European whitefish (*Coregonus lavaretus*), primarily a freshwater species, the proteomes of hatch-stage samples from 27 families in two populations and five salinity treatments were analysed. Several proteins were identified to play key roles in osmoregulation, most importantly a highly conserved cytokine, tumour necrosis factor; whereas calcium receptor activities were associated with salinity adaptation (Papakostas *et al.* 2012).

Conclusion

Environmental proteomics has established itself as a powerful approach to investigate the cellular effects of environmental stresses. It investigates and identifies the proteins associated with multiple biological processes, intracellular signalling pathways and key proteins that are potentially involved in the regulation of stress adaptation in aquatic organisms. Thus it would allow exploring possible approaches to mitigate the more pronounced biological effects of these environemental stressors.

References

Abe, H., Hirai, S. & Okada, S., 2007. Metabolic responses and arginine kinase expression under hypoxic stress of the kuruma prawn Marsupenaeus japonicus. Comparative Biochemistry and Physiology - A Molecular and Integrative Physiology. https://doi.org/10.1016/j.cbpa.2006.08.027

Bonga, S.E.W., 1997. The stress response in fish. Physiological reviews 77(3), 591-625.

Bosworth, C.A., Chou, C.W., Cole, R.B. & Rees, B.B., 2005. Protein expression patterns in zebrafish skeletal muscle: initial characterization and the effects of hypoxic exposure. Proteomics 5(5), 1362-1371.

Chen, K, Cole, R.B. & Rees, B.B., 2013. Hypoxia-induced changes in the zebrafish (Danio rerio) skeletal muscle proteome. Journal of Proteomics 78, 477-485.

Diaz, R.J. & Rosenberg, R., 1995. Marine benthic hypoxia: a review of its ecological effects and the behavioural responses of benthic macrofauna. Oceanography and Marine Biology: An Annual Review. Vol. 33.

Donaldson, E.M., 1981. The pituitary-interrenal axis as an indicator of stress in fish. Stress and fish, pp.11-47.

Dowd, W.W., Renshaw, G.M.C., Cech, J.J. & Kultz, D., 2010. Compensatory proteome adjustments imply tissue-specific structural and metabolic reorganization following episodic hypoxia or anoxia in the epaulette shark (Hemiscyllium ocellatum). Physiol. Genomics 42(1), 93–114.

Hochachka, P.W. & Lutz, P.L., 2001. Mechanism, origin, and evolution of anoxia tolerance in animals. In Comparative Biochemistry and Physiology - B Biochemistry and Molecular Biology. https://doi.org/10.1016/S1096-4959(01)00408-0

Ibarz, A., Martín-Pérez, M., Blasco, J., Bellido, D., de Oliveira, E. & Fernández-Borràs, J., 2010. Gilthead sea bream liver proteome altered at low temperatures by oxidative stress. Proteomics 10, 963–975.

IPCC. 2018. Global Warming of 1.5 °C. IPCC Special Report 1.5 - Summary for Policymakers. IPCC. https://doi.org/10.1017/CBO9781107415324

Iwama, G.K., 1998. Stress in fish. Annals of the New York Academy of Sciences 851(1), 304-310.

Ky, C.L., De Lorgeril, J., Hirtz, C., Sommerer, N., Rossignol, M. & Bonhomme, F., 2007. The effect of environmental salinity on the proteome of the sea bass (Dicentrarchus labrax L.). Animal Genetics 38(6), 601-608.

Madeira, C., Mendonça, V., Leal, M.C., Flores, A.A.V., Cabral, H.N., Diniz, M.S. & Vinagre, C., 2017. Thermal stress, thermal safety margins and acclimation capacity in tropical shallow waters—An experimental approach testing multiple end-points in two common fish. Ecological Indicators. https://doi.org/10.1016/j.ecolind.2017.05.050

Mahanty, A., Purohit, G.K., Banerjee, S., Karunakaran, D., Mohanty, S. & Mohanty, B.P., 2016. Proteomic changes in the liver of Channa striatus in response to high temperature stress. Electrophoresis. https://doi.org/10.1002/elps.201500393

McLean, L., Young, I.S., Doherty, M.K., Robertson, D.H.L., Cossins, A.R., Gracey, A.Y., Beynon, R.J. & Whitfield P.D., 2007. Global cooling: Cold acclimation and the expression of soluble proteins in carp skeletal muscle. Proteomics 7, 2667–2681.

Nuez-Ortín, W.G., Carter, C.G., Nichols, P.D., Cooke, I.R. & Wilson, R., 2018. Liver proteome response of pre-harvest Atlantic salmon following exposure to elevated temperature. BMC Genomics. https://doi.org/10.1186/s12864-018-4517-0

Oehlers, L.P., Perez, A.N. & Walter, R.B., 2007. Detection of hypoxia-related proteins in medaka (Oryzias latipes) brain tissue by difference gel electrophoresis and de novo sequencing of 4-sulfophenyl isothiocyanate-derivatized peptides by matrix-assisted laser desorption/ionization time-of-flight mass spectrometry. Comp Biochem Physiol C Toxicol Pharmacol. 145(1), 120-133.

Papakostas, S., Vasemägi, A., Vähä, J.P., Himberg, M., Peil, L. & Primmer, C.R., 2012. A proteomics approach reveals divergent molecular responses to salinity in populations of European whitefish (Coregonus lavaretus). Molecular Ecology. doi: 10.1111/j.1365-294X.2012.05553.x

Pickering, A.D. (ed.), 1981. Stress and fish (No. 04; QL615, P5.). London: Academic Press.

Randall, D.J., & Perry, S.F., 1992. Catecholamines. In W. S. Hoar and D. J. Randall (eds.), Fish physiology, Vol. 12B, pp. 255–300. Academic Press, New York.

Reid, S.G., Bernier, N.J. & Perry, S.F., 1998. The adrenergic stress response in fish: control of catecholamine storage and release. Comp Biochem Physiol C Toxicol Pharmacol. 120(1), 1-27.

Smith, R.W., Cash, P., Ellefsen, S. & Nilsson, G.E., 2009. Proteomic changes in the crucian carp brain during exposure to anoxia. Proteomics 9(8), 2217-2229.

Wedemeyer, G.A., Barton, B.A. & McLeay, D.J., 1990. Stress and acclimation. In C. B. Schreck and P. B. Moyle (eds.), Methods for fish biology, pp. 451–489. American Fisheries Society, Bethesda, Maryland.

Wedemeyer, G.A. & McLeay, D.J., 1981. Methods for determining the tolerance of fishes to environmental stressors. Stress and fish, pp. 247-275.

29

Introduction to R-statistics

D K Karna

Introduction

R is a language and environment for statistical computing and graphics. It is a GNU project which is similar to the S language and environment which was developed at Bell Laboratories (formerly AT&T, now Lucent Technologies) by **John Chambers and colleagues**. R can be considered as a different implementation of S. There are some important differences, but much code written for S runs unaltered under R.

R provides a wide variety of statistical (linear and nonlinear modelling, classical statistical tests, time-series analysis, classification, clustering) and graphical techniques, and is highly extensible. The S language is often the vehicle of choice for research in statistical methodology, and R provides an Open Source route to participation in that activity. In a nutshell, R is an open-source software environment for statistical computing that is rapidly becoming the tool of choice for data analysis in the life sciences and elsewhere. It's developed by a large international community of scientists and programmers and is at the forefront of new developments in statistical computing. Home page: http://www.r-project.org/

R is an integrated suite of software facilities for data manipulation, calculation and graphical display. It includes

- an effective data handling and storage facility,
- a suite of operators for calculations on arrays, in particular matrices,
- a large, coherent, integrated collection of intermediate tools for data analysis,
- graphical facilities for data analysis and display either on-screen or on hardcopy, and
- a well-developed, simple and effective programming language which includes conditionals, loops, user-defined recursive functions and input and output facilities.

- R is not a menu driven software and has a steep learning curve. Generally, it is operated through command line interface though some menu-driven packages are developed in R, those are not very popular

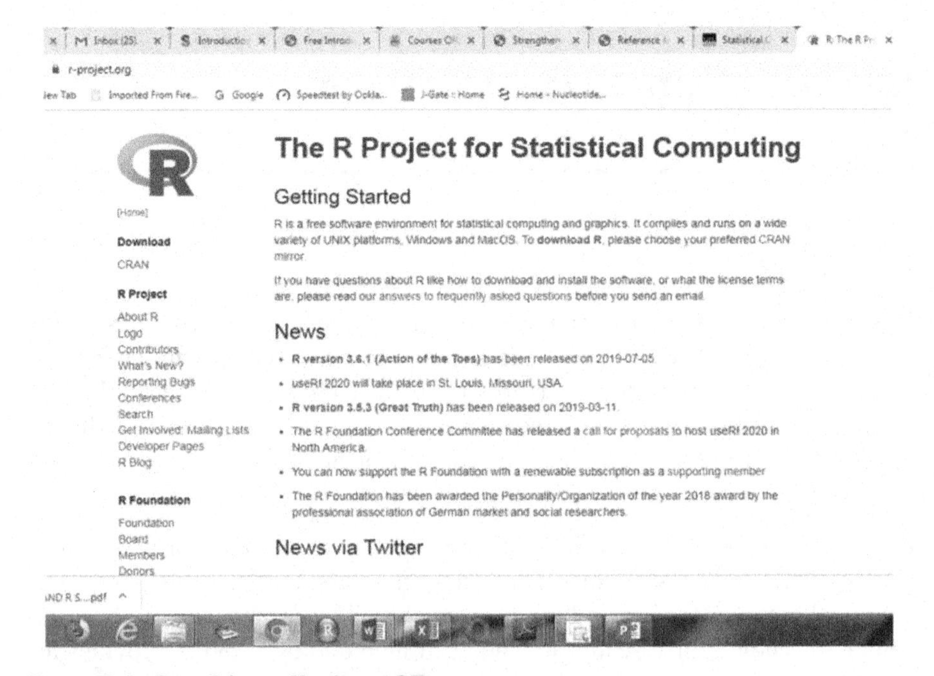

Downlolad and installation of R

R is freely available in the CRAN sites (Comprehensive R Archive Network) which can be visited from the links of R home page or through site: https://cran.r-project.org/mirrors.html. R software compiles and runs on a wide variety of UNIX platforms, Windows and MacOS. The latest version of R: R version 3.6.1 for Windows (Action of the Toes) has been released on 5th July 2019 which is available both for 32 bit and 64 bit processors. To download R one has to chose the version as per the OS he is using and chose a mirror site on CRAN Network. The installation can simply be carried by running the installer R-3.6.1-win.exe and it finishes smoothly.

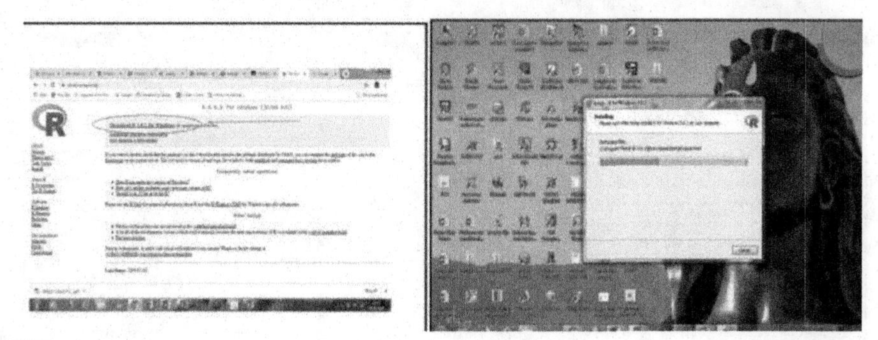

The R interface

Once the R application is run, it produces a simple interface with a console.

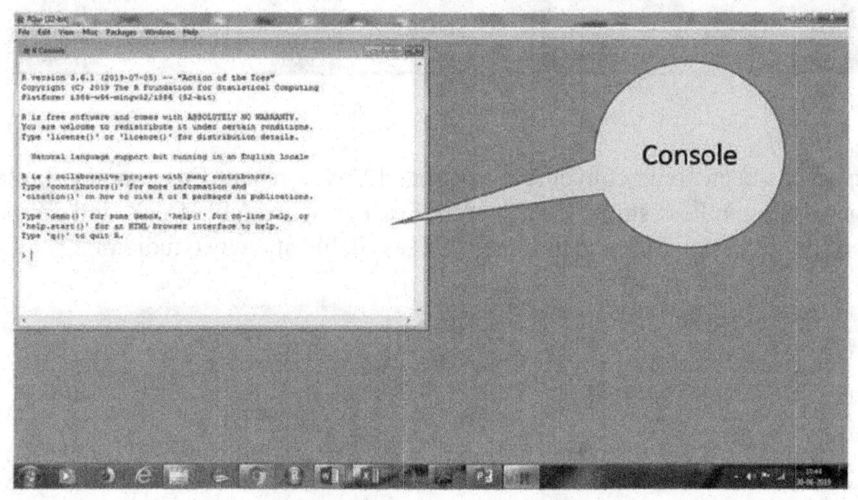

It is the window where commands are written to R and executed. It is almost a blank screed with a cursor indicated by “>” followed by blip. The result is seen immediately followed by the command.

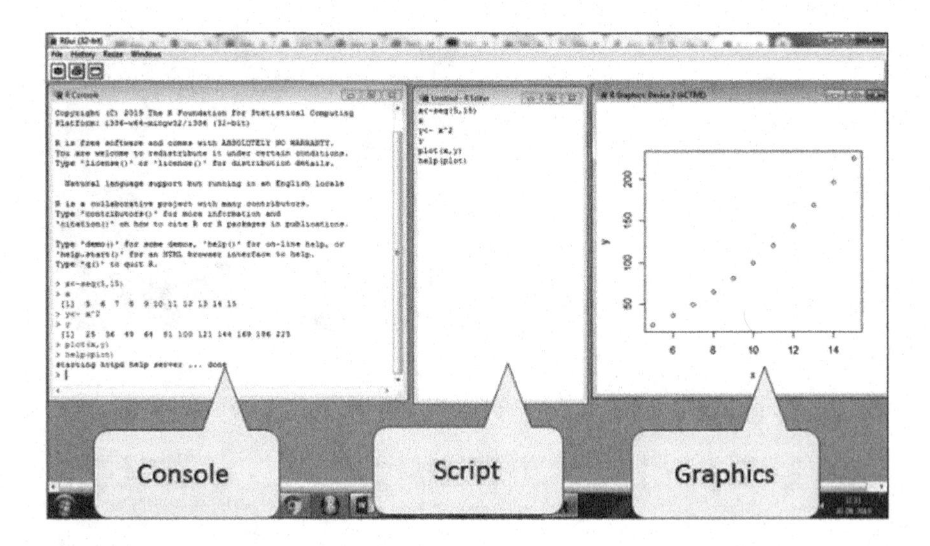

R Studio

There has been some convenient Integrated Development Environment build for working in R. R studio is such an IDE which is available freely and can be used doing R statistics and graphics. It is available at www.rstudio.org.

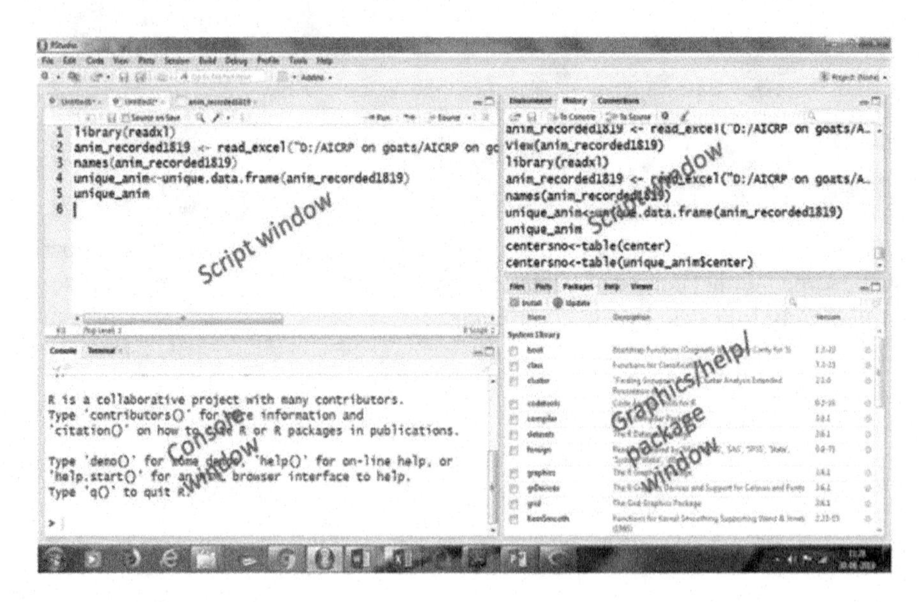

Basic Operations in R

The scripts for performing basic operations are as follows:

```
#IMPORTANT NOTE FOR WINDOWS USERS:
#R gets confused if you use a path in your code like
#c:\mydocuments\myfile.txt
#This is because R sees "\" as an escape character. Instead, use
#c:\\my documents\\myfile.txt
#c:/mydocuments/myfile.txt# print the current working directory - cwd
# change to mydirectory
setwd("c:/Rwork")
getwd()
# list the objects in the current workspace
ls()
x<-9
y<-12
x^15
z<x*y
z<-x*y
z
p<-12*x +13*y
p
#creating a vector in R
x<-c(1,2,3,5,6,7)
x
#sequence function to get a series of data
x<-seq(1:50)
x
y<-seq(12)
```

```
y<-x^2
y
q<- 24
x>y
x
y
z<-runif(10, min=10,max=20)
#to get other information about runif command
help(runif)
#creating vector by concatenating
x<-c(1,3,4,5,7)
#combining vectors to make dataset
z
#to count the number of the elements
length(x)
length(y)
x+y
x-y
x*y
log(x)
exp(x)
#R comes with a number of sample datasets that you can experiment with.
#Type data ( ) to see the available datasets. The results will depend on
#which packages you have loaded. Type help(datasetname) for details
#on a sample dataset.
data ()
help(ChickWeight)
ChickWeight
```

```
#viewing of data to see the first five line and last five line of the dataset
head(dat.csv)
tail(dat.csv)#to know the data structure
str(dat.csv)
# to know the data dimension >dim(dataset)
dim(dat.csv)
```

Descriptive statistics using R

The scripts command line are given as follows for getting the descriptive statistics

from the body of data.

```
## Descriptive Statistics using R
read.csv("test.csv", header =TRUE)
d<-read.csv("test.csv", header =TRUE)
d
#1 mean,median,25th and 75th quartiles,min,max
summary(d)
#2 # n, nmiss, unique, mean, 5,10,25,50,75,90,95th percentiles
# 5 lowest and 5 highest scores
library(Hmisc)
describe(d)
#3 using pastecs package
# nbr.val, nbr.null, nbr.na, min max, range, sum,
# median, mean, SE.mean, CI.mean, var, std.dev, coef.var
library(pastecs)
stat.desc(d)
#4 Using Psych package
# item name ,item number, nvalid, mean, sd,
# median, mad, min, max, skew, kurtosis, se
library(psych)
```

```
describe(d)
#5 summary statistics by groupng variables
describeBy(d,d$race)
describeBy(d,d$female)
```

Graphics using R

The scripts are given for plotting graphs from the data as follows along with the comments.

```
#R CODES FOR GRAPHICS
#to know what is the working directory
getwd()
#to assign the working directory
setwd("D:/Rwork")
getwd()
#to open excel file
install.packages("openslsx")
install.packages("xlsx")
install.packages("gdata")
library(openxlsx)
#to open the excel file
students<-read.xlsx("D:/Rwork/students.xlsx")
students<-read.csv("students1819.csv")
#to view first five rows of data with all the columns
head(students)
students[1:5,]
#to view first four columns of data with all rows
students[,1:4]
dim(students)
str(students)
```

```
student1<-students[,c(3,6,7,8,9)]

student1

attach(students)

names(students)

#to get histogram

hist(BW)

#to give title, x axix label

hist(BW, main=”Variation in Body Weights of students”, xlab=”Body
Weigthts in kg”)

#to get number of boxes, x axis limit, y axis limit)

hist(BW, main=”Variation in Body Weights of students”, xlab=”Body
Weigthts in kg”,

breaks =9, xlim=c(0,90),ylim=c(0,10) )

hist(BW, main=”Variation in Body Weights of students”, xlab=”Body
Weigthts in kg”,

breaks =9, xlim=c(45,100),ylim=c(0,10) )

#to change the color of the box to red 2,green 3,blue 4, 5,6,7,etc

hist(BW, main=”Variation in Body Weights of students”, xlab=”Body
Weigthts in kg”,

breaks =5, xlim=c(45,100),ylim=c(0,10) , col=6)

#to get the lines overlaid over the histogrAM

lines(density(BW))

#to change the color and thickness of the lines

lines(density(BW), col=2,lwd=4)

#to get help to make box plot from the data

#to get the help on the barplot

help(barplot)

#alternatively

?barplot
```

```
#to producefrequency table
#table You can generate frequency tables using the table( ) function,
#tables of proportions using the prop.table( ) function,
#and marginal frequencies using margin.table( ).
## 2-Way Frequency Table
##attach(mydata)
##mytable<- table(A,B) # A will be rows, B will be columns
##mytable # print table
##margin.table(mytable, 1) # A frequencies (summed over B)
##margin.table(mytable, 2) # B frequencies (summed over A)
##prop.table(mytable) # cell percentages
##prop.table(mytable, 1) # row percentages
##prop.table(mytable, 2) # column percentages
#to get the relative frequency
summary(students)
table(students$SEX)
t<-table(students$SEX)
frq<-(table(students$SEX))/22
frq
barplot(t)
#to produce barplotof the frequency
barplot(frq)
#to give title to the graph,label the x axis and y axis
barplot(frq,main=”Male Female Ratio”,
xlab=”Gender”,ylab=”Percentage”)
#to set the values of y axis aligned parallel
barplot(frq,main=”Male Female Ratio”,
xlab=”Gender”,ylab=”Percentage”, las=1)
```

```
#to change the labels appearing under each of the bar
barplot(frq,main=”Male Female Ratio”,xlab=”Gender”,
ylab=”Percentage”, las=1,names.arg = c(“Females”,”Males”))
#to make the bars horizontal
barplot(frq,main=”Male Female Ratio”,ylab=”Gender”,
xlab=”Percentage”, las=1,names.arg = c(“Females”,”Males”),
horiz=TRUE)
##to produce stacked bar chart
#to see the relationship between the categorical variables
table1<-table(SEX,MAJOR)
table1
#to produce bar plotone stacked over other
barplot(table1)
#to produce the bar plot one beside the other
barplot(table1,beside=T)
#to produce the default legend by the side
barplot(table1,beside=T, legend.text=T)
#to give customised legend and adding a title
barplot(table1,beside=T, legend.text=c(“Girls”, “Boys”),
main=”Body Weights of males and females as per stream”)
#to change the color of the bar and axis labels
barplot(table1,beside=T, legend.text=c(“Girls”, “Boys”),
main=”Body Weights of males and females as per stream”,
col=c(2,3), xlab=”Subject Streams”, ylab=”Body Weights”,las=1)
#set the limit of the axis
barplot(table1,beside=T, legend.text=c(“Girls”, “Boys”),
main=”Body Weights of males and females as per stream”,
col=c(2,3), xlab=”Subject Streams”, ylab=”Body Weights”,las=1,
```

```
ylim=c(0,6),border=5)
##to pproducebarplot of students in different majors
count<-table(MAJOR)
count
sum(count)
##to get freq in percentages
pc<-(count/sum(count))*100
pct<-format(round(pc,2),nsmall=2)
pct
#to produce barplot
barplot(count,main=”number students in different MAJOR”,col = c (1,2,3,
4,5,6,7))
## to produce the Barplot for the mean body weight of boys and girls
head(students)
mean.bw<-tapply(BW,SEX,mean)
n.bw<-tapply(BW, SEX,length)
sd.bw<-tapply(BW,SEX,sd)
sem<-sd.bw/sqrt(n.bw)
mean.bw
sem
mids<-barplot(mean.bw,xlab=”Sex groups”, ylab=”Body Weight of
students”,ylim=c(0,80))
arrows(mids,mean.bw-sem,mids,mean.bw+sem,code=3,angle=90,length=0.3)
text(mids,2,paste(“n=”,n.bw))
##to produce mosaic plot
??mosaic
mosaicplot(table1)
mosaicplot(table1,color=T)
mosaicplot(table1,color=c(2,4))
```

```
table2<-table(MAJOR,SEX)
mosaicplot(table2)
mosaicplot(table2, color=T)
mosaicplot(table2,color=c(2,4))
#to make a pie chart
#to get help to produce a pie
help(pie)
pie(frq)
pie(t)
#to add title
pie(frq,main=”Chart indicating male female
proportions”,col=c(“red”,”blue”))
#to put the chartinside the box
box()
##to get the pie chart on majors
pie(count)
labls<-paste(MAJOR,pct,”%”,sep=” “)
pie(count,labels =labls )
length(count)
##to produce 3D exploded pie chart
library(plotrix)
pie3D(count,labels=labls,explode=0.4,
main=”Pie Chart of majors”)
help(boxplot)
?boxplot
# to make box plot for variable
boxplot(BW)
# to confirm that
```

```
quantile(BW, probs = c(0,0.25,0.5,0.75,1))
# to give a title tot the box plot and putting Y axis label
boxplot(BW, main="Boxplot of Body Weights", ylab="Body Weights")
#to change color of title,axis,grpahs,labelscol.main,col.axis,
#col.lab or col can be used
boxplot(BW, col=3,main="Boxplot of Body Weights", ylab="Body
Weights",col.main=2)
#To change the style of axis label las set to 1 horizontal
boxplot(BW, col=3,main="Boxplot of Body Weights", ylab="Body
Weights",col.main=2, las=3)
##to parallel
boxplot(BW, col=3,main="Boxplot of Body Weights", ylab="Body
Weights",col.main=2, las=1)
#to make boxplot by different class variables like sex
boxplot(BW~sex,col=3,main="Box plot by Gender")
boxplot(BW~SEX,col=3,main="Box plot by Gender")
# To create boxplot by subsetting of groups
boxplot(BW[SEX=="female"],BW[SEX=="male"],col=7,border=2)
head(students)
## to produce scatterplot from the data
plot(AGE,HEIGHT)
#to give title,change plot symbols,colors and rotate text
plot(AGE,HEIGHT, main="Scatterplot of Age vs Height",
pch=5,col=2,las=1)
#to change the size of the plotting character
plot(AGE,HEIGHT, main="Scatterplot of Age vs Height",cex=2,
pch=5,col=2,las=1)
# to change the limit of the axis
plot(AGE,HEIGHT, main="Scatterplot of Age vs Height",cex=2,
pch=8,col=2,
```

```
las=1, xlim=c(22,30),ylim=c(50,80))
#to get a linear regression line for predicting Height from Age
abline(lm(HEIGHT~AGE))
#to change the color,type and size of the line
abline(lm(HEIGHT~AGE), col=4,lty=2,lwd=6))
###to explore the help of available plotting parameters
help(par)
```

30

Application of Basic Statistics for Analysis for Climate Related Data and Designing Experiment for Measurement of Climatic Effect

G D Nayak, K K Sardar and B C Das

Introduction

Statistics generally refers to information about an activity or a process that is expressed in quantitate as well as qualitative traits. The quantitative research process is well structured whereas the qualitative one is fairly unstructured. It depends upon how a piece of information have been collected and analyzed. A good researchers need to have both types of skill. This article highlights about the operational procedure, about methods, procedures and techniques that are used in analysis of climate related data.

Parameter vs Statistic

When a data is collected with survey of all the individuals (like in a census) and analysed for certain values, it is called a parameter. But when, we guess the same value with a random sample and interpret for whole population it is called a statistic. Environmental parameters are: mean, variance, correlation, regression, heritability, repeatability etc. Again, there are two types of variables: quantitative and qualitative. Quantitative variables can take any value in a specified range. So, they are of continuous type. But qualitative variables are very few types (2 ,3 or 4 types) and they are of discrete type. Body weight is an example of quantitative variable whereas, comb type is a qualitative variable.

Central tendency values

This is such a value that describes the characteristics of entire mass of data. One example is Arithmetic mean. But sometimes this becomes insufficient to describe a mass of data. For that matter, we think of spread of data along with the central values that describes the data mass in a more defined manner. Variance is such a value that represents the presence of uniformity for a character in a population. Greater the uniformity, smaller is the variance value for that character.

Variance= $\sigma^2 = \sum(X - \mu)^2 / N$

Where, X = individual value, μ = Arithmetic men of the data set, N = total number of observations.

∑ = summing up of all square of the mean deviated individual values

Standard deviation = $\sqrt{(\text{variance})} = \sigma$

To compare the variability two or more series of data, we calculate coefficient of variation.

Coefficient of variation = C.V. = (σ / μ) x mean

When two cows are observed for daily milk yield for 6 consecutive days, the Coefficient of variation is calculated for each of the cow. The cow having more C.V. is less consistent.

Standard error

We must have seen tables in research article where in a mean value is followed by a ± sign and then a value is written thereafter. This after value is about the error associated with your calculation. It is customary to write Mean ± S.E values.

Standard error = S.E. = $\sigma / \sqrt{n}$ where, σ= standard deviation, n = number of observations.

Standard error gives an idea about unreliability of a sample and it gives the limits within which the parameter values are expected to lie. So, 1.53±0.03 litres of average milk yield means our data is expected to lie between 1.53 to 1.47 litres of milk yield.

Correlation

When we study relationship between two variables (ex. Body weight at 6 weeks age and life time egg production value of poultry), the value that measures this relationship is done through calculation of correlation coefficient (r).

$r = [(\sum X.Y) - (\sum X).(\sum Y)/N)] / [(\sum X^2) - ((\sum X)^2/N)] \times [(\sum Y^2) - ((\sum Y)^2/N)]$

$= \text{covariance}_{xy} / \sqrt{(\text{variance}_x \text{ X variance}_y)}$

Example :

Poultry No.	X (life time egg number)	Y (day old bodyweight, grams)
1.	100	30
2	200	40
3.	50	20
4.	100	35

In this small hypothetical example, $\sum X.Y = (100 \times 30) + (200 \times 40)..........+ (100 \times 35) = 15{,}500$

$\sum X = 100 + 200 +......+ 100 = 450$

$\sum Y = 30 + 40 + ...+ 35 = 125$

$N = 4$

$\sum X^2 = 100^2 + 200^2 + ...+ 100^2 = 62{,}500$

$\sum Y^2 = 30^2 + 40^2 + ...+ 35^2 = 4125$

$r = [(15{,}500) - ((450 \times 125)/4))] / \sqrt{[((62{,}500) - ((450)^2/4))] \times [(4125) - ((125)^2/4)]}$

$= 0.89$

The value of r shall always lie between +1 and -1. When r is a positive number decimal, we call it a positive correlation and a negative decimal value, it is a negative correlation. A positive correlation means one variable increases in tandem with increase in other variable. A negative correlation indicates a decrease in value of one variable when the other variable value increases. If r=correlation coefficient, then r^2 value is called coefficient of determination. For example, r=0.9, then r^2=0.81. i.e. 81% of variation in one variable has been due to other variable.

Regression

Some times when we see two variables, one happens to be a dependant variable and other is an independent one. One example can be cited here is, temperature and milk yield. Suppose we have average maximum temperature of 12 months and also average milk yield of a farm for those 12 months. Surely we can find r value. But here we can see that, temperature is not dependent on milk yield, rather milk yield is dependent on temperature. So, here average maximum temperature is an independent variable, whereas, milk yield is a dependent variable. Again based on regression coefficient value we can predict or estimate the amount of milk production on 13 th month for that farm when the average maximum temperature is say, 35^0 C.

The regression line equation is

Y = a + b X where, y = dependent variable, X= independent variable, a= intercept and b= regression coefficient.

b_{YX}=[(∑X.Y)- ((∑X).(∑Y)/N)]/ [(∑x^2)- ((∑X)^2/N)] = covariance $_{xy}$ / variance$_x$

a= (Mean of Y) – b.(mean of X)

Now, predicted Y value = a + b . (a given X value)

Test of hypothesis

While conducting an experiment, we take a sample of data and based on this sample data we give inference about a population parameter. So, before conducting an experiment we have to begin with an assumption called a hypothesis. The assumption is usually about a population parameter like, mean, variance, etc. Actually in statistics, we set up two different types of hypothesis: (a) Null hypothesis (H_0) and (b) Alternative hypothesis (H_1). Null hypothesis is about equality of parameters of different samples i.e. there is no difference between these sample in the parameter value. The alternative hypothesis is about inequality of parameters of different samples i.e. there exists a difference among the samples in the parameter value.

Steps in hypothesis testing

i) First set up a hypothesis about a population parameter.

ii) Setting a test: There are several tests like Z-test, t-test, F-test, chi-square test etc available for different types of data. We the researcher has to identify which type of test is needed for a particular set of data analysis and inference.

iii) A calculation with a given formula is carried out with the sample data. This is called test statistic value.

iv) Setting up of a suitable significance level: Usually two types of significance level are used in statistics. They are 1% level of significance and 5% level of significance. In a very simplistic manner, we can say that, significance at 1% means our inference can go wrong in 1 case out of 100 cases. Similarly, 5% significance means, we can go wrong in 5 cases out of 100 cases.

v) If test statistic is greater than or equal to t table value at n-1 degrees of freedom at a chosen level of significance, null hypothesis (H0) is rejected. That means alternative hypothesis is accepted. That means, all parametric values are not equal.

vi) If test statistic is less than t table value at n-1 degrees of freedom at a chosen level of significance, null hypothesis (H0) is accepted. That means alternative hypothesis is rejected. That means, all parametric values are equal and there is no difference between them.

vii) Making decisions.

t-test

It is mostly used for small sized samples. There are several types of t-test are in use. But here we shall describe only the two important t-tests usually carried out by researchers.

One sample t-test

Here we have to test whether a random sample drawn from the population is a giving any idea about the population? One example can suffice to explain this phenomenon.

A random sample of buffaloes of size 16 has a mean of 53 liters of monthly milk yield. The population of buffaloes has a mean monthly milk yield of 42 liters. Test whether, the sample yield is actually representing the population or not.

Data on milk yield of 16 buffaloes :

1. 40

2. 38

3. 58

4. .

5. .

6. .

16.65

The standard deviation value i.e. $\sigma = 2.53$ (suppose) using these 16 data given above.

i. H_0: $\mu = \mu_0$ H_1: $\mu \neq \mu_0$ where, μ = mean of sample and μ_0 = mean of population.

ii. One sample t-test is to be carried out.

iii. t_c = t value calculated =[(mean of sample)- (mean of population)]√n / σ and here modulus value i.e. only the positive value has to be considered.

So, t_c =[53-42]√16 /2.53

Or t_c = 11 x 4 /2.53 = 18.3

iv. let us compare this value with a t-table value (usually given in the last pages of any statistics book) at n-1 degrees of freedom i.e. 16-1=15 d.f. at 5% level of significance.

v. Suppose, the t table value at 15 d.f. with 5% level of significance is 2.83.

vi. Now, $t_c > t_{tab}$ i.e. 18.3 > 2.83.

vii. H_0 rejected.

viii. The sample mean is not equal to the population mean. Then we can infere that the sample that is taken randomly does not belong to the population,

ix. In a similar way, we can have an inference at 1% level of significance also.

Two sample t-test

It is usually carried out to test the difference between means of two samples.

Here, $H_0 : \mu_1 = \mu_2$ and $H_1 : \mu_1 \neq \mu_2$

t_c = t value calculated=[[(mean of one sample)-(mean of other sample)]√[(n_1. n_2)/ (n_{1+} n_2)]]/ S

where, n_1 is equal to the number of observations in sample 1 and n_2 is equal to the number of observations in sample 2.

S = √ [[(n_1-1)s_1^2 + (n_2-1)s_2^2]/ (n_{1+} n_2-2)] where s_1^2 is the variance of sample 1 and s_2^2 is the variance of sample 2.

Now, comparison of t-table value with t- calculated value at chosen level of significance and n_{1+} n_2-2 degrees of freedom, is carried out and inference is given as usual described in previous sections.

Example

Two types of drugs are used on 5 and 7 goats respectively for increasing their body weight at 1 year age. The increase in weight for each is given below :

Drug A : 14, 11,13,12,,10

Drug B : 9,10,15,14,12,9,8

We can find out easily t_c value which can be compared with t-table value at $n_{1+} n_2$-2 degrees of freedom under 5% as well as 1% significance level and accordingly inference can be given for this experiment.

Analysis of variance

Analysis of variance in short is called as ANOVA. It is a statistical technique specially designed to test the means (or for that matter a parameter) of more than two samples for equality or inequality. Two types of ANOVA we shall discuss here although several types are available there in statistical books.

i. One – way classification (otherwise called as CRD, i.e. completely randomized design)

ii. Two- way classification (otherwise called as RBD, i.e. randomized block design)

Here two terms that are most important are: treatment and replication. Treatments are a set of conditions imposed on the experimental unit (animals) usually of environmental in nature. Replications are repetitions of a single treatment. A typical ANOVA table is attached here with each kind of design. The common terms in ANOVA are, degrees of freedom (d.f.), Sum of squares (S.S), Mean squares (M.S.) and F calculated (F_c). Calculation of these items for each source of variation can be seen from any type of experimental design book available in the market.

CRD

It will be easy and precise to explain all the process of CRD through a suitable example that is described below:

In a poultry feeding experiment, 4 different types of feeds (with different levels of protein) were fed to 5 birds each . The data were recorded on their 6th week body weight in kg.The experiment started at 2nd week age of the birds and continued up to 7th week age. The aim of the experiment was to see whether different level of protein affects the body weight of poultry birds at 6th week age or not.

Feed 1 : the feed have 22% protein in the feed mixture = treatment 1=t_1

Feed 2 : the feed have 19% protein in the feed mixture = treatment 2=t_2

Feed 3 : the feed have 15% protein in the feed mixture = treatment 3=t_3

Feed 4 : the feed have 25% protein in the feed mixture = treatment =t_4

Five birds were taken randomly from the poultry house and allocated with feed 1 feeding. Similarly, another 5 birds were taken randomly from the poultry house and allocated with feed 2 feeding. Again, another 5 birds were taken randomly from the poultry house and allocated with feed 3 feeding and at last another 5 birds were taken randomly from the poultry house and allocated with feed 4 feeding. All the birds were of same age group. Here the treatment, i.e. t is equal to 4 and replication i.e. r is equal to 5. We have to test whether mean body weight at 6 week age with different type of feeds are same or not.

Here, $H_0 : \mu_1 = \mu_2 = \mu_3 = \mu_4$ and $H_1 : \mu_1 \neq \mu_2 \neq \mu_3 \neq \mu_4$

Detail data sheet is given below

	r_1	r_2	r_3	r_4	r_5
t_1	2	3	2	1	2
t_2	5	1	2	1	3
t_3	4	4	3	1	2
t_4	5	2	2	2	3

ANOVA						
Source of Variation	*SS*	*df*	*MS*	*F*	*P-value*	*F crit*
1. Between treatments	2.2	3	0.733333	0.437811	0.729015	3.238872
2.Within treatments (Error)	26.8	16	1.675			
3. Total	29	19				

F_c =F value calculated through ANOVA = 0.43

F_t=F value found in the table =3.238872 at 5% significance level

$F_c < F_t$ at 5% level of significance. So, the feeds have no effect on 6th week body weight of poultry in this experiment. All the means under each treatment is same. Had it been a case of $F_c > F_t$, we would have drawn an interpretation that, different feeds have significant effect on 6th week body weight of poultry. Similar comparisons and interpretations can also be made at 1% level of significance also.

After significance test, if the treatments are found significantly affecting the data (trait) then go for C.D.(critical difference) value calculation and compare difference in two treatment values with the C.D. value.

RBD

It is a better experimental design in comparison to CRD. In CRD we had only one source of variation, that is feed, affecting each data. But in RBD, we have two sources of variation affecting each data. Let us take one example .

There are 4 locations based on a different THI (Temperature-humidity-Index) values. Five breeds of lactating cattle are used in this experiment. So, here, breeds are treatments and locations are blocks (B1, B2 etc). Thus five treatments and 4 blocks are used in this experiment. The data is recorded on daily milk yield of each cattle.

	B1	B2	B3	B4
T1	2	4	3	8
T2	3	3	9	5
T3	5	6	7	9
T4	3	3	7	1
T5	1	1	8	2

ANOVA						
Source of Variation	*SS*	*df*	*MS*	*F*	*P-value*	*F crit*
Treatments	34.5	4	8.625	1.7753	0.198613	3.259167
Blocks	48.2	3	16.06667	3.307033	0.057385	3.490295
Error	58.3	12	4.858333			
Total	141	19				

F_c =F value calculated through ANOVA = 1.77

F_t=F value found in the table =3.25 at 5% significance level

$F_c < F_t$. at 5% level of significance. So, the feeds have no effect on lactational yield in this experiment. All the means under each treatment is same. Had it been a case of $F_c > F_t$, we would have drawn an interpretation that, different breeds have significant effect on lactational yield. Similar comparisons and interpretations can also be made at 1% level of significance also.

After significance test, if the treatments are found significantly affecting the data (trait) then go for C.D.(critical difference) value calculation and compare difference in two treatment values with the C.D. value.

Here one thing is clear that the error that is caused in the data due to locations is extracted out and then F value were calculated which more valid for comparison than the CRD way.

There are certain designs which can quantify a particular environmental effect. Some designs are there for calculating genotype environment interaction effects. After a handful knowledge of environmental effects, we can certainly rear with a certain kind of environment or we may take out such environments in order to have better production.

31

Climate Change and Policy Interventions for Livestock Sector

Gopal Krushna Tripathy

Introduction

In India, almost 85 % of our population resides in rural areas and the same proportion is dependent on agriculture for sustenance and animal husbandry for additional income. Distribution of livestock is more equitable compared to that of land and about 85 percent of livestock are owned by the landless, marginal and small landholding families. This sector provides employment for the farmers through livestock farming as well as in processing, value addition and marketing of livestock products.

Rapid population, economic growth, increased demand for livestock products such as meat and dairy products, influences the present livestock production systemtowards intensive production. There is increased conflict over scarce resources such as land, water, food grains etc. Losing livestock assets for rural communities might lead to abject poverty with serious effect ontheir livelihoods. The climate change impact will worsen the vulnerability of livestock systems and those farmers depending on livestock, particularly the poor.

Sensitivity of livestock production to climate change

Climate change is a phenomenon being experienced by the mankind since its origin on the earth. Due to impact of Green House Gas causing higher ambient temperatures, as well as occurrence of natural disasters such as cyclone, floods and droughts, are likely to adversely affect poultry and livestock. Higher temperatures would limit the feed intake of poultry birds, which will reduce the growth of chicken, broilers and other birds. Many empirical study reveals that rising temperature is likely to lower the productivity potential of crossbred animals. The rising temperature also decreases the total dry matter intake and milk yield in indigenous cows. The more hot–humid climatic conditions and the extending summer would plausibly aggravate the heat stress in animals and further adversely affect their productive and reproductive performance. Compared to other species of domestic animals, poultry is more sensitive to high ambient temperatures as they do not have sweat glands, have a higher basal metabolic rate per unit body weight and body is covered with feathers. With further increase in atmospherictemperature due to climate change in future, the poultry birds will face more stress. Already depleted grazing lands would be less productive due to rising salinity in coastal areas and droughts. Higher temperatures and humidity may affect animal health through the more rapid breeding of parasites and bacteria. The performance (e.g., growth, milk and wool production, reproduction), health would be affected by climate change. These changes are likely to seriously affect the livelihoods of livestock farmers and the availability of livestock products in India.

Policy Measures needed for Livestock Sector to adapt and mitigate climate change effect

Climate change resulting from accelerated pace of release of Green-house Gases (GHGs) is the major issue. Livestock production system is directly challenged due to expected rise in temperature, changes in quantum & pattern of rainfall, annual frequency of occurrence of extreme weather events.

Livestock production system is sensitive to climate change and at the same time itself a contributor to the phenomenon, climate change has the potential to be an increasingly formidable challenge to the development of the livestock sector in India. This sector has a very important role to play in the economic progress of the country as it contributes over one-fourth (26%) to the agricultural GDP and provides employment to nearly 20 million people in directly or indirectly. Global demand for livestock products is expected to double by 2050, mainly due to improvement in the worldwide standard of living. Meanwhile, climate change is a threat to livestock production because of the impact on quality of feed crop and forage, water availability, animal and milk production, livestock diseases, animal reproduction, and bio- diversity.

Prime Minister Council of India has formulated National Action Plan on Climate Change in 2008. There are eight missions and out of which National Mission on sustainable agriculture is more relevant to livestock sector. Paris climate conference in 2015 achieved a universal agreement in climate with the aim of keeping global warming below 2^0C above pre-industrial levels.

The livestock development strategy in the changing climate scenario should essentially focus on minimization of potential production losses resulting from climate change, on one hand, and on the other, intensify efforts for Green House Gas (GHG)reduction from livestock sector as this would also be instrumental in increasing production of milk by reducing energy loss from the animals through methane emissions. Responding to the challenge of climate change requires formulation of appropriate policy that support and facilitate the implementation of climate change adaptation and mitigation measures. Climate change mitigation and adaptation policies for livestock sector will generally achieve greater acceptance, if they can enhance productive efficiency, raise farmers' incomes, and reduce food costs. Hence, there is strong need for research that lower GHG emissions and produce economic and environmental co-benefits. Some of the policy options based on various published articles are outlined for consideration.

1. Foster the adoption of cost effective technology and management practices for reducing methane generation and emission from livestock farming by modification of diet and methane management.
2. Develop capacity for disaster management planning at the local level preparedness planning, vulnerability mapping while preparing the community level disaster management plans, in livestock and dairy sectors.
3. Promote Integrated Farming in rainfed agriculture farming system. Livestock (both large & small ruminants) should form essential components of integrated farming, as they are more climate resilient than crop systems. ICAR has developed large number of Integrated farming models suitable to different agro-climatic situations, which may be adopted as suitable to local context.
4. Promote pasture, silvi-pasture systems, fodder trees, multiple tree based systems in non-arable lands, particularly in village common lands to increase carbon sequestration and encourageimproved grazing management methods increase soil organic carbon content.
5. In view of climate change implications causing rising temperatures and rainfall deviations, develop and promote the varieties of fodder crop cultivar that are tolerant to heat and moisture stress in different regions.
6. As majority of livestock owners are resource poor, economic instruments in the form of subsidies may be applied across sections, and include a variety of policy design such as financial incentives and microfinance, grants, loans and institutional credit with low interest rate are required. The institutional support in the form of livestock insurance, disaster contingency fund should be made available.
7. The preference of the farmers should also be taken into consideration for prioritization of breeds. Research on inherent capabilities of different indigenous breeds and identification of characteristics that can better adapt to climatic condition.

8. Livestock breeding policy specific to different agro- ecological zones need to be designed to ensure that farmers have access to the best animals in each environment. Genetic improvement of breeds needs to focus on identifying high-performance indigenous elite breeds and the optimisation of their potential for multiple functions, rather than on a single productive trait such as meat or milk production.
9. Indian dairy sector is characterized more by 'production by masses' than 'mass production' and the milk productivity in the country remains low. Despite of low productivity level, technology led productivity enhancement has been important in contributing to the growth in milk production. The growth in milk yield has been particularly pronounced in case of local cows. Therefore, due attention on the productivity enhancement of the indigenous cows through genetic improvement by using elite indigenous breeds will be a win-win strategy, as indigenous cows have good adaptation potential to climate stress compared to buffaloes and crossbreds.
10. Livestock breeding policy must enable in-situ conservation by involving livestock owners in long-term sustained initiatives. This ensures focus on conservation and development of indigenous breeds, along with protection of pastoralists communities, who are the real breeders.
11. Improve fertility in dairy cattle which would require less replacement of breeding animals and in turn decreases the green house gas.
12. Livestock schemes need to be more suited to dry-land farming systems. Crossbreds deliver high milk yields, generating better economic returns in the short run. However, given the impacts of climate change, such as increased dry spells, droughts, high temperatures during summers, acrossbred dairy farming does not seem to be a viable option.
13. Shift the approach from high economic returns to more ecosystem-friendly sustainable modes of production. The animal husbandry and rural development programmes must support conservation and development of indigenous breeds by providing subsidies/ incentives or loan schemes to encourage farmers to rear them.
14. Mixed farming systems which existed earlier need to be promoted for integrating livestock in the circular bio-economy (output from one system becomes input for other) and for ensuring the development and conservation of cattle in optimal numbers, coupled with conservation and development of Common Property Resources to support the extensive system of animal husbandry.

15. Small livestock is essential for the rural poor to cope with the emerging risks caused by climate change. For a comprehensive and holistic development of sheep, goat, pig and poultry in the climate change scenario, it is also essential that the bio-resources, that contribute to feed and fodder basket are tapped. There is a need for a special programme for backyard desi poultry, which provides important subsidiary income for women, nutrition for poor households, and plays an important role in traditional rituals and cultural activities in rural India.
16. Sheep and goat farming in potential areas need to be supported under Government schemes to help the poor for adapting climate change risk. Priority should be given to the provision of grazing rights, particularly to pastoralist communities, and to accommodate grazing based livestock production systems to protect and regenerate common property resources.
17. Give focus for quick diffusion of cutting edge climate technology through collaborative research and development for such future technologies. Research on heat stress effect, implications of reduced water availability for livestock system due to climate change, non-rumnant production system,marker assisted selection incorporating traits pertaining to production, adaptation and low methane emission in breeding program are some of the important areas of concern. Establish green research fund for stregthening research work.
18. Develop and design mitigation policies to improve information about the linkage between production practices and technologies and emissions. Technology subsidies can reduce emissions by shifting production from polluting methods of production, toward cleaner methods.
19. Incentives to farmers to integrate forage production with food and other crops. Improving feed quality can be achieved through improved grassland management, improved pasture species (e.g. grass and legumes mix), forage mix, feed processing (e.g. chopping, urea treatment) and strategic use of supplements, preferably locally available.
20. Reducing the incidence and impact of diseases, parasites and insect burdens will result in higher productivity and efficiency through a strong disease surveillance system.
21. Fodder cultivation and feed mixtures should receive due attention to support livestock rearing. In addition to cultivating fodder on the farms, the gochars (land reserved as a common land of the village for fodder) should be secured and rejuvinated to use as pasture land. Improvement in quantity and quality of straws and stovers along with grain in breeding

programs of cereal food crops would ensure availability of feed for livestock.

22. Climate resilient breeds using advanced bio-technology tools and identify ideal genotype traits to quantify heat stress response to livestock. Incentives to farmers for active participation in such participatory technology development need to be provided.
23. Subsidize the cost to producers of using an emission-reducing technology by providing incentives for installation of biogas plants for efficient use of dung for clean fuel and manure as well as waste to compost programme.
24. IEC program for facilitating farmers to cope with climate change and its impacts should be taken up by strengthening the extension mechanism. Setting up of demonstration units for climate resilient livestock production system and technologies with government support for sensitising the farmers is essential.
25. Farmers need to be educated for scientifically designed shelters and improved management of animals including health management for livestock suitable and acceptable indifferent bioclimatic zones to minimize the effects of climate and climate change on livestock production.
26. Training in animal climatology needs to be strengthened/ increased to take up the research work climate change area at major research institutes / colleges.

Conclusion

Free and unrestricted access to the atmosphere, which has a limited capacity to absorb solar radiation, has resulted in excessive warming. Few individuals contribute for emitting GHG or destroys carbon 'sinks', while the costs of these actions are shared by everyone. Collective action through a comprehensive government policy is required to reduce GHG emissions.

Small, marginal farmers and landless labourers own almost 70 percent of livestock in India and the animals of these resource poor livestock owners are more prone to climate change and are at greater risk. Limiting the effects of climate change for those vulnerable groups is necessary to achieve sustainable development and equity, including poverty eradication.

The crop livestock system is one of the most important characteristics of Indian agrarian economy and livestock sector is the integral part of India's agriculture sector. Indian livestock sector provides sustainability and stability to the national economy by contributing to farm energy and food security. Livestock sector not only provides essential protein and nutrition to human diet through milk, eggs, meat and by products such as hides and skin, blood, bone and fat etc., but also plays an important role in utilization of non-edible agricultural by-products. At the same time, livestock has a major influence on the environment through its effects on water and air quality, deforestation and biodiversity. Livestock sector also contribute to greenhouse gas (GHG) emissions. There is a substantial potential to reduce the sector's contribution to climate change through policies that foster the adoption of a wide range of technologies and management that are available to reduce emissions from livestock farming. Incentive-based policies should be used to encourage the adoption and diffusion of cost effective technical options to reduce GHG emission. The implications for technology transfer is that and 'push' achieved through stronger external support will be effective only if complemented by the 'pull' of stable policy framework that sustain demand at the receiving end.

The climate change action plan of India was formulated during the year 2008, which is now obsolete. There is a pressing need to revisit the National Action Plan on Climate Change and to reformulate comprehensive climate policy in India. Policy integration, institutional design for effective implementation, and climate justice must play a central role. Livestock sector being a State subject, each State should have a suitable policyframeworkalong with enhanced investment in developmental schemes/ programs in livestock sector for mitigating and adapting the possible climate change effectare critical to protect livestock production. Livestock is a cross-cutting sector involving Agriculture, Water, Rural Development, Renewable Energy, Health, Environment Ministries etc. and coordination is a big challenge.

References

Aydinalp, C., Cresser, M.S., 2008: The effects of climate change on agriculture. Agric. Environ. Sci. 5: 672–676.

FAO paper, Livestock solutions for climate change

M. Melissa Rojas-Downing, A. PouyanNejadhashemi, Timothy Harrigan, Sean A. Woznicki: Climate change and livestock: Impacts, adaptation, and mitigation

National Academy of Agricultural Sciences, New Delhi May, 2016 - Climate Resilient Livestock Production

P.K. Pankaj, D.B.V. Ramana, R. Pourouchottamane and S. Naskar; Livestock Management under Changing Climate Scenario in India

Report of the Committee on Doubling Farmers' Income; Volume XIV, Ministry of Agriculture & Farmers Welfare

S.M.K. Naqvi and V. Sejian; Global Climate Change, 2011; Role of Livestock

Smita Sirohi & Axel Michaelowa, 2007; Sufferer and cause: Indian livestock and climate change.

WOTR-Livestock Position Paper, Evolving Practical Strategies for Adaptation to Climate Change.

Printed in the United States
by Baker & Taylor Publisher Services